Nanohertz Gravitational Wave Astronomy

Nanohertz Gravitational Wave Astronomy

Stephen R. Taylor

CRC Press
Taylor & Francis Group
Boca Raton London New York

CRC Press is an imprint of the
Taylor & Francis Group, an **informa** business

First edition published 2022
by CRC Press
6000 Broken Sound Parkway NW, Suite 300, Boca Raton, FL 33487-2742

and by CRC Press
2 Park Square, Milton Park, Abingdon, Oxon, OX14 4RN

Library of Congress Cataloging-in-Publication Data

Names: Taylor, Stephen R., author.
Title: Nanohertz gravitational wave astronomy / Stephen R. Taylor.
Description: First edition. | Boca Raton, FL : CRC Press, 2022. | Includes bibliographical references and index.
Identifiers: LCCN 2021028439 | ISBN 9781032147062 (hardback) | ISBN 9780367768621 (paperback) | ISBN 9781003240648 (ebook)
Subjects: LCSH: Gravitational waves. | Astronomy.
Classification: LCC QC179 .T39 2022 | DDC 523.01/9754--dc23
LC record available at https://lccn.loc.gov/2021028439

ISBN: 978-1-032-14706-2 (hbk)
ISBN: 978-0-367-76862-1 (pbk)
ISBN: 978-1-003-24064-8 (ebk)

DOI: 10.1201/9781003240648

Typeset in LM Roman
by KnowledgeWorks Global Ltd.

Contents

Preface

Gravitational waves are a radically new way to peer into the darkest depths of the cosmos. Almost a century passed from their first prediction by Albert Einstein (as a consequence of his dynamic, warping space-time description of gravity) until their direction detection by the LIGO experiment. This was a century filled with theoretical and experimental leaps, culminating in the measurement of two black holes inspiraling and merging, releasing huge quantities of energy in the form of gravitational waves.

However, there were signs along the way to this first detection. Pulsars are rapidly rotating neutron stars that emit radiation along their magnetic field axes, which may be askew from their rotation axis. This creates a lighthouse effect when the radiation is swept into our line of sight. Through tireless observations of the radio pulse arrival times, we are able to construct detailed models of the pulsar's rotation, binary dynamics, and interstellar environment. The first hint of gravitational waves came from the measured orbital decay of a binary star system that contained a pulsar. The decay was in extraordinary agreement with predictions based on gravitational-wave emission.

In fact, pulsars can be used to make direct detections of gravitational waves using a similar principle as LIGO. When a gravitational wave passes between a pulsar and the Earth, it stretches and squeezes the intermediate space-time, leading to deviations of the measured pulse arrival times away from model expectations. Combining the data from many Galactic pulsars can corroborate such a signal, and enhance its detection significance. This technique is known as a Pulsar Timing Array (PTA). PTAs in North America, Europe, and Australia have been active for the last couple of decades, monitoring almost one hundred ultra-stable pulsars, with the goal of measuring gravitational waves entering the Galaxy from inspiraling supermassive black-hole binary systems at cosmological distances. These black-hole systems are the most massive compact objects in the Universe, having masses almost a billions times as big as our Sun, and typically lurk in the hearts of massive galaxies. They only form binary systems when their host galaxies collide together.

In this book, I provide an overview of PTAs as a precision gravitational-wave detection instrument, then review the types of signal and noise processes that we encounter in typical pulsar data analysis. I take a pragmatic approach, illustrating how our searches are performed in real life, and where possible directing the reader to codes or techniques that they can explore for themselves. The goal of this book is to provide theoretical background and practical recipes for data exploration that allow the reader to join in the exciting hunt for very low frequency gravitational waves.

I would not have been able to write this book without the continued collaboration with many excellent colleagues in NANOGrav and the International Pulsar Timing Array. Some have directly assisted in reviewing early chapter drafts, and I am particularly grateful to Dr. Joseph Romano, Dr. Xavier Siemens, Dr. Michele Vallisneri, Dr. Katerina Chatzioannou, Dr. Daniel D'Orazio, and Mr. William Lamb for this reason. More practically, I would not have had the time to write this book without the overwhelming patience and support from my incredible wife, Erika, who tolerated many Saturday and Sunday afternoons of me shut away while I prepared this. Our very cute cat Olive also kept me company during many writing sessions. All my family back in Northern Ireland continue to be a source of strength and encouragement for me.

Stephen R. Taylor
Nashville, Tennessee
May 2021

About the Author

Stephen R. Taylor is an Assistant Professor of Physics & Astronomy at Vanderbilt University in Nashville, Tennessee. Born and raised in Lisburn, Northern Ireland, he went on to read Physics at Jesus College, Oxford, England from 2006–2010, before switching to the "other place" for his PhD from the Institute of Astronomy at the University of Cambridge in 2014. His positions have included a NASA Postdoctoral Fellowship at NASA's Jet Propulsion Laboratory, and a NANOGrav Senior Postdoctoral Fellowship at the California Institute of Technology in Pasadena, California. He currently lives in Nashville with his wife Erika and cat Olive.

CHAPTER 1

A Window onto the Warped Universe

Homo Sapiens' split from *Neanderthals* and other human species occurred roughly 500,000 years ago, while the cognitive revolution that gifted us a fictive language took place 70,000 years ago. It has been argued that this fictive language allowed us to develop abstract modes of thinking and planning that set us apart from other species. For most of the history of our species on this planet, the practice of *"Astronomy"* has involved staring at the night sky to divine meaning. Indeed, when humanity first watched the twinkling fires in the sky, they envisioned a rich canvas on which their myths and collective stories took place. The heavens were the realm of the gods, and seeing patterns in the positions of stars may have been humanity's first attempt to make sense out of a complex Universe. The human eye was the most sensitive astronomical instrument for the vast majority of our time on Earth. Yet the history of Astronomy, much like the history of the human species, is one of widening panoramas.

The last 500 years has seen an explosion in creativity, knowledge, technology, art, and humanist values. Specifically to astronomy, the late 16th and early 17th centuries brought us the humble optical telescope, with which Galileo Galilei made precision observations of our Solar System that included the rings of Saturn and the four largest moons of Jupiter. His simple observation of those moons orbiting Jupiter caused a revolution, seeming to contradict Aristotelian cosmology. Yet reason and knowledge prevailed, eventually compelling society to change its worldview based on observations.

Progress accelerates even more rapidly in the 19th century, when scientists learn that electromagnetic waves can have wavelengths greater and smaller than what the human eye can perceive. In fact, humans are innately blind to the vast majority of the electromagnetic spectrum! William Herschel, discoverer of Uranus, was the first to probe the spectrum beyond visible light in 1800. Through his investigation of the wavelength distribution of stellar

DOI: 10.1201/9781003240648-1

spectra, Herschel discovered infrared radiation. Only one year after this, Johann Ritter discovered ultraviolet radiation, followed in 1886 by Heinrich Hertz's discovery of radio waves, then microwaves. In 1895, Wilhelm Roentgen revolutionizes diagnostic medicine with his discovery of X-rays that can travel through the human body. Finally, in 1900, Paul Villard finds a new type of radiation in the radioactive emission from radium, which is later ascribed to be a new, even shorter wavelength form of electromagnetic radiation called gamma-rays. See how the rate of progress accelerates, having taken us hundreds of thousands of years to find infrared radiation, yet less than a century to fill in the rest of the spectrum.

These other parts of the electromagnetic-wave landscape were used throughout the 20th century to probe hitherto unexplored vistas of the cosmos. Infrared radiation has longer wavelengths than visible light and can pass through regions of dust and gas with less absorption, enabling the study of the origin of galaxies, stars, and planets. Ultraviolet radiation allows us to study the properties of young stars, since they emit most of their radiation at these wavelengths. Radio waves can easily propagate through the Earth's atmosphere, allowing observations even on a cloudy day. The impact of radio observations on astronomy is too great to chronicle here, but it has a particularly special relevance to this book, which will be explored later. Most people have microwave emitters in their homes to heat up food, yet the same type of radiation bathes the cosmos in its influence as a record of when the Universe was only 380,000 years old – this is the Cosmic Microwave Background that has revolutionized precision cosmology. X-ray astronomy has allowed us to peer into highly energetic environments like supernovae and the accretion flows that surround neutron stars or black holes. Finally, the class of objects known as gamma-ray bursts are the incredibly energetic and explosive result of the implosion of high mass stars and the collision of neutron stars. The latter is a literal treasure trove; double neutron star mergers are likely one of the environments that forged all the heavy elements (like gold, platinum, and silver) found on Earth. These collisions are the result of the inexorable inspiral of binary orbits due to the energy loss to gravitational wave emission.

Just as electromagnetic waves are oscillations in the electromagnetic field caused by charge acceleration, gravitational waves are oscillating solutions to the space-time metric of Einstein's geometric theory of gravity in the presence of accelerating massive objects. They were predicted almost as soon as Einstein had developed his General Theory of Relativity (GR) in 1915, yet it was a full century before theory and technology could advance enough to detect them. In that century, Einstein would denounce them as coordinate artifacts, physicists would ignore them as mathematical curiosities that could not be practically measured, some pioneers would claim detections that turned out to be false flags, the first evidence of their existence would be seen in the orbital decay of binary pulsars, and a new millennium would arrive, all before the detection of two colliding black holes in 2015.

But, what are these specters of spacetime known as gravitational waves? We can not "see" them, nor perceive them in the slightest with any human sense. They travel at the speed of light, deforming spacetime as they propagate, and interacting very weakly with matter along the way. Their presence is exceptionally difficult to infer. In fact, the LIGO instrument that detected them first needed to be able to measure length deviations equivalent to the width of a human hair over the distance between Earth and Alpha Centauri, about 3 lightyears. That's a length deviation of about 1 part in 10^{21}; an almost unfathomably miniscule departure from perfection, and yet a departure that enables a fundamentally new form of astronomy.

Measuring gravitational waves (GWs) allows us to study systems from which no light is emitted at all. Pairs of black holes with no gas surrounding them will be completely invisible to conventional electromagnetic astronomy, yet will resound loudly in GWs. Black holes are extreme solutions to Einstein's theory, and are the crucible for successor theories to be tested that may link GR to a quantized description of gravity. Mining catalogs of GW signals will unearth clues about star formation, the progenitor environments of black holes and neutron stars, and even allow us to measure the rate of expansion of the Universe. As I write, Earth-based GW detectors like LIGO and Virgo have detected more than 50 signals from inspiraling pairs of black holes and neutron stars. Such detectors are limited in sensitivity to GW frequencies of ~ 10 Hz $\leq f \leq 1$ kHz. But just as the electromagnetic landscape was ripe for discovery beyond the visible spectrum, so too is the GW landscape, where the most titanic black holes in the Universe lie in wait at the heart of massive galaxies for Galaxy-scale GW detectors to find them. Let us take a look at how precision timing of rapidly spinning neutron stars from all across the Galaxy will chart this "undiscovered country" of the low-frequency warped Universe.

CHAPTER 2

Gravity & Gravitational Waves

2.1 GRAVITY BEFORE AND AFTER EINSTEIN

Gravity is the dominant dynamical influence throughout the Universe. Yet pick a ball off the ground, and the modest mechanical pull of your muscles has beaten the combined gravity of an entire planet.

2.1.1 Standing on the Shoulders of Giants

The Renaissance saw humanity climbing slowly out of the darkness and ignorance of the middle ages, and shedding their worshipful attitude to the knowledge of their ancient forebears. To question the cosmological wisdom of Aristole and Ptolemy was stupid and possibly even heretical! Nevertheless, in the 16th century, the Polish astronomer Nicholaus Copernicus proposed a new heliocentric model of the Universe. This was later put to the test using the observations of Denmark's Tycho Brahe, who compiled high-accuracy stellar catalogues over more than 20 years of observations at his observatory, Uraniborg. Johannes Kepler, who had been Brahe's assistant, used these data to derive his famous laws of planetary motion. In his 1609 book *Astronomia nova*, Kepler presented two of these laws based on precision observations of the orbit of Mars, and in so doing ushered in the revolutionary new heliocentric model of the solar system with elliptical, rather than circular, planetary orbits.

Arguably the greatest scientific breakthrough of this age came in 1687, when Sir Isaac Newton published the *Principia* (1). Ironically (given that the author of this book is writing it during the COVID-19 pandemic) Newton first glimpsed the concept of a universal law of gravitation 20 years prior while under self-quarantine at his family's farm during a bubonic plague outbreak. Although it had been separately appreciated by contemporaries such as

DOI: 10.1201/9781003240648-2

Hooke[1], Wren, and Halley, Newton's demonstrations of the universal inverse-square law of gravitation were remarkable not only for their ability to derive Kepler's laws, but most crucially for their predictive accuracy of planetary motion. Simply put, this law states that the mutual attraction between two *bodies* is proportional to the product of their masses, and the inverse square of their separation. When applied to the motions of solar system bodies it was extraordinarily successful, and remains so today—it guided Apollo astronauts to the Moon and back. The mark of a successful theory is the ability to explain new observations where previous theories fall short; in 1821, the French astronomer Bouvard used Newton's theory to publish prediction tables of Uranus' position that were later found to deviate from observations. This discrepancy motivated Bouvard to predict a new eighth planet in the solar system that was perturbing Uranus' motion through gravitational interactions. Independent efforts were made by astronomers Le Verrier (of France) and Adams (of England) to use celestial mechanics to compute the properties of such an additional body. Le Verrier finished first, sending his predictions to Galle of Berlin Observatory, who, along with d'Arrest, discovered Neptune on September 23, 1846.

Neptune's discovery was a supreme validation of Newtonian celestial mechanics. But the underlying theory of gravity suffered two fundamental flaws: it did not propose a mediator to transmit the gravitational influence, and it assumed the influence was instantaneous. Furthermore, Le Verrier, who had wrestled with complex celestial mechanics calculations for months, had by 1859 noticed a peculiar excess precession of Mercury's orbit around the Sun. Various mechanisms to explain this were proposed, amongst which was an idea drawn from the success of Neptune; perhaps there was a new innermost planet (named *Vulcan*) that acted as a perturber on Mercury's orbit. A concerted effort was made to find this alleged new planet, and there were many phony claims by amateur astronomers that were taken seriously by Le Verrier. But all attempts to verify sightings ultimately ended in failure. This problem would require a new way of thinking entirely, and a complete overhaul of the more than 200 years of Newtonian gravity.

2.1.2 The Happiest Thought

Newton's universal law of gravitation treated gravity as a force like any other within his laws of mechanics. Albert Einstein had already generalized Newtonian mechanics to arbitrary speeds $v \leq c$ in his 1905 breakthrough in special relativity, and went about doing the same for gravity. His eureka moment came, as it usually did for Einstein, in the form of a *Gedankenexperiment* (a thought experiment). Imagine an unfortunate person falling from a roof; ignoring wind resistance and other influences, this person is temporarily weightless

---

[1]In fact, the title of this sub-section is a phrase used by Newton in reference to Hooke, which some have interpreted as a thinly-veiled jab at the latter's diminutive height.

and can not *feel* gravity. A similar restatement of this concept is that if we were to place a sleeping person inside a rocket without windows or other outside indicators, and accelerated the rocket upwards at 9.81 ms^{-2}, then upon waking this person would feel a downwards pull so akin to Earth-surface gravity that they could not tell the difference. These thought experiments are consequences of the *Equivalence Principle*, which states that experiments in a sufficiently small freely-falling laboratory, and carried out over a sufficiently short time, will give results that are identical to the same experiments carried out in empty space.

In 1907, Hermann Minkowski developed an elegant geometric formalism for special relativity that recast the theory in a four-dimensional unification of space and time, unsurprisingly known as "spacetime". Even today we refer to spatially flat four-dimensional spacetime as *Minkowski spacetime*. Accelerating particles are represented by curved paths through spacetime, and with the thought-experiment equivalence principle breakthrough that allowed Einstein to see acceleration as (locally) equivalent to gravitational influence, he began to build connections between geometric spacetime curvature and the manifestation of gravity. It was a long road with many blind alleys that required Einstein to learn the relevant mathematics from his friend and colleague Marcel Grossman. But finally, in late 1915, Einstein published his general theory of relativity (GR)[2]. Formulated within Riemannian geometry, Einstein's new theory described how spacetime geoemtry could be represented by a metric tensor, and how energy and momentum leads to the deformation and curvature of spacetime, causing bodies to follow curved free-fall paths known as geodesics. In essence, gravity is not a "pull" caused by a force-field; it is the by-product of bodies traveling along free-fall paths in curved spacetime. The shortest possible path between two points remains a straight line (of sorts), but for the Earth or other planets, this straight-as-possible path within the curved geometry created by the Sun is one that carries it around in an orbit. In searching for a pithy summary, it's difficult to beat Wheeler: *"Space-time tells matter how to move; matter tells space-time how to curve"*.

The phenomenal success of this new paradigm, connecting relativistic electrodynamics with gravitation, overturning notions of static space and time, and demonstrating gravity to emerge from the curvature of the fabric of spacetime, can not be understated. The first real demonstration of its power came through being able to predict the excess perihelion precession of Mercury's orbit that had been observed by Le Verrier and others; in GR this appears as a conservative effect due to the non-closure of orbits in the theory. After computing the correct result, Einstein was stunned and unable to work for days afterward. But this was a post-hoc correction of a known observational curiosity. The real test of any scientific theory is being able to make testable, falsifiable predictions.

---

[2]Interesting discussions in November/December 1915 between Einstein and the mathematician David Hilbert have raised some questions of priority, but Einstein undeniably developed the breakthrough vision of gravity as geometry.

Einstein next set his sights on the deflection of starlight passing close to the Sun. Newtonian gravity does predict that such deflection will happen at some level, but precision observation of such deflection provided a perfect testbed for the new curved spacetime description of gravity. Before he had fully fleshed out his theory, Einstein made a prediction in 1913 that he encouraged astronomers to test. Before they could do so, World War I broke out, making the unrestricted travel of anyone to test a scientific theory a somewhat dangerous prospect. There is a serendipity to this though – his calculation was wrong! It would only be using the final form of his field equations that Einstein made a testable prediction for the deflection of starlight by the Sun that corresponds to twice the Newtonian value. This was later confirmed in 1919 by the renowned astronomer Sir Arthur Eddington (already a proponent of Einstein's work) who made an expedition to the isle of Principe for the total solar eclipse on May 29th of that year. The eclipse made it possible to view stars whose light grazes close to the surface of the Sun, which would otherwise have been completely obscured. This result made the front page of the New York Times, catapulting Einstein from a renowned figure in academia to a celebrity scientist overnight.

Since those early days, GR has passed a huge number of precision tests (2), and sits in the pantheon of scientific theories as one of the twin pillars of 20th century physics, alongside quantum theory. In 1916, Einstein proposed three tests of his theory, now dubbed the "classical tests". Perihelion precession and light deflection were the first two of these, having been validated in very short order after the development of the theory. Gravitational redshift of light (the third and final classical test) would take several more decades, eventually being validated in 1960 through the *Pound–Rebka Experiment* (3). The experiment confirmed that light should be blueshifted as it propagates toward the source of spacetime curvature (Earth), and by contrast redshifted when propagating away from it.

Beyond merely deflection, GR predicts that light will undergo a time delay as it propagates through the curved spacetime of a massive body. By performing Doppler tracking of the *Cassini* spacecraft en route to Saturn, this *Shapiro delay* effect (4) was experimentally tested, with the data agreeing to within 10^{-3} percent of the GR prediction (5). The 2004 launch of the *Gravity Probe B* satellite afforded independent verifications of the GR effects of *geodetic* and *Lense-Thirring precession* of a gyroscope's axis of rotation in the presence of the rotating Earth's curved space-time, exhibiting agreements to within 0.28% and 19%, respectively (6). Furthermore, there is a whiff of irony in the fact that Newtonian gravity was used to compute the trajectory of Apollo spacecraft to the Moon, yet one of the greatest legacies of the Apollo project is the positioning of retro-reflectors on the lunar surface that enable laser ranging tests of the *Nordtvedt effect* (an effect which, if observed, would indicate violation of the strong equivalence principle), *geodetic precession*, etc. (7). A more exotic and extraordinary laboratory for testing the strong equivalence principle is through the pulsar hierarchical triple system J0337+1715 (8), composed of an inner neutron star and white dwarf binary, and an outer white dwarf. The outer white dwarf's gravitational interaction with the inner binary causes the pulsar and its white dwarf companion to accelerate, but their response to this (despite their differences in compactness) differs by no more than 2.6×10^{-6} (9).

Many more experiments have since validated GR. Those most relevant to the subject of this volume are through the indirect and direct detection of gravitational waves from systems of compact objects. These will be discussed in more detail in the next section.

2.2 GRAVITATIONAL WAVES

2.2.1 A Brief History of Doubt

Gravitational waves (GWs) have had a colorful history. They were first mentioned by Henri Poincare in a 1905 article that summarized his theory of relativity, and which proposed gravity being transmitted by an *onde gravi-fique* (gravitational wave) (10). In Einstein's 1915 general theory of relativity, an accelerating body can source ripples of gravitational influence via deformations in dynamic space-time (11; 12). Whereas electromagnetic waves are sourced at leading order by second time derivatives of dipole moments in a charge distribution, mass has no "negative charge" and thus GWs are instead sourced at leading order by second time derivatives of the quadrupole moment of a mass distribution.

Whether they even existed at all or were merely mathematical curiosities was the subject of much early speculation. By 1922, Sir Arthur Eddington (who had conducted the famous solar eclipse experiment that tested Einstein's light deflection prediction), found that several of the wave solutions that Einstein had found could propagate at any speed, and were thus coordinate artifacts. This led to him quipping that GWs propagated "*at the speed of thought*" (13). But the most famous challenge to the theory of GWs came from Einstein himself, who, in collaboration with Nathan Rosen, wrote an article in 1936 that concluded the non-existence of GWs. This chapter in the history of GWs is infamous, because when Einstein sent his article along to the journal *Physical Review*, he became angry and indignant at receiving referee comments, stating in response to editor John Tate that "*we had sent you our manuscript for publication and had not authorized you to show it to specialists before it is printed*" (14). Einstein withdrew his article. After Rosen departed to the Soviet Union for a position, Einstein's new assistant Leopold Infeld befriended the article's reviewer, Howard Robertson, and managed to convince Einstein of its erroneous conclusions arising from the chosen cylindrical coordinate system. The article was heavily re-written, the conclusions flipped, and eventually published in *The Journal of the Franklin Institute* (15).

The thorny issue of coordinate systems was solved two decades later by Felix Pirani in 1956, who recast the problem in terms of the Riemann curvature tensor to show that a GW would move particles back and forth as it passed by (16). This brought clarity to the key question in gravity at the time: whether GWs could carry energy. At the first "GR" conference held at the University of North Carolina at Chapel Hill in 1957, Richard Feynman developed a thought-experiment known as *"the sticky bead argument"*. This stated that if a GW passed a rod with beads on it, and moved the beads, then the motion of the beads on the rod would generate friction and heat. The passing GWs had thus performed work. Hermann Bondi, taking this argument and Pirani's work on the Riemann curvature tensor, fleshed out the ideas into a formal theoretical argument for the reality of GWs (17).

Bolstered by the theoretical existence of GWs, Joseph Weber established the field of experimental GW astronomy by developing resonant mass instruments (also known as *Weber bars*) for their detection. By 1969/1970, he made claims that signals were regularly being detected from the center of the Milky Way (18; 19), however the frequency and amplitude of these alleged signals were problematic on theoretical grounds, and other groups failed to replicate his observations with their own Weber bars. By the late 1960s and early 1970s, astronomy was swept up in the excitement of the discovery of *pulsars*. These rapidly rotating neutron stars acted like cosmic lighthouses as they swept beams of radiation around, allowing astronomers to record and predict radio-pulse arrival times with extraordinary precision. In 1974, Russell Hulse and Joseph Taylor discovered B1913+16, the first ever pulsar in a binary system (20). Over the next decade, the observed timing characteristics of this pulsar reflected its motion alongside the companion, allowing the binary system itself to be profiled. As shown in Fig. 2.1, the system exhibits a shift in the time to periastron corresponding to a decay in the orbital period, which matches to within 0.16% the loss in energy and angular-momentum predicted by GR as a result of GW emission (21). The discovery of this pulsar binary system garnered the 1993 Nobel Prize in Physics. Further precision binary-pulsar constraints on GW emission have been made possible by the double-pulsar binary system J0737−3039 (which constrains the GR prediction to within 0.05% (22)), and the pulsar–white-dwarf system J0348+0432 (23).

The early 1970s also saw the origins of research into detecting GWs through laser interferometry, a concept that posed significant technological challenges. The foundations of GW interferometry were laid by Forward (25), Weiss (26), and others, which motivated the construction of several highly successful prototype instruments. These early detectors had arm lengths from ∼10 m to ∼100 m, and reached a strain sensitivity of ∼10^{-18} for millisecond burst signals, illustrating that the technology and techniques were mature enough to warrant the construction of much longer baseline instruments. Over decades of research, this concept bloomed into the Laser Interferometer Gravitational-wave Observatory (LIGO) in the USA, which became an officially-funded National Science Foundation project in 1994. LIGO is a project that includes two instruments: LIGO-Hanford (in Washington), and LIGO-Livingston (in Louisiana). Other instruments have been built since, including the Virgo

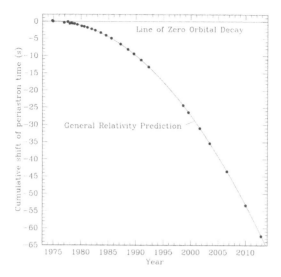

Figure 2.1: Data and GR prediction of the cumulative shift of periastron time of pulsar B1913+16, showing extraordinary agreement with orbital decay through GW emission. Figure reproduced with permission from Ref. (24).

detector in Italy, and now the KAGRA instrument in Japan. There are plans to continue to expand this global network of laser interferometers in order to drastically improve the sky localization of detected signals, and thus narrow search regions for electromagnetic follow-up.

At this stage, history begins to impinge on the modern day, and I'll defer further discussions of LIGO until after introducing the theoretical framework of GWs. The last piece of history that I'll discuss requires us to jump forward to 2014, when the BICEP2 collaboration announced an incredibly exciting result (27). It appeared that a polarization pattern had been found in the cosmic microwave background that could be consistent with so-called B-modes (or curl modes). These modes could have been imprinted from quantum fluctuations in the gravitational field of the early Universe that had been inflated to cosmological scales. While the raw data was excellent, unfortunately such primordial GWs were not the only potential source of B-modes; galactic dust, if not properly modeled, could impart this polarization pattern as well. Subsequent analysis now suggests that dust is the main culprit, and primordial GW signatures are much smaller than initially thought (e.g., 28). However, the hunt is still on for this elusive imprint from the dawn of time through later generations of BICEP and other detectors.

2.2.2 Waves from Geometry

Let's dig into some illuminating mathematics to grasp what these gravitational waves (GWs) really are. General Relativity is enshrined within the *Einstein Field Equations*,

$$G_{\mu\nu} := R_{\mu\nu} - \frac{1}{2}Rg_{\mu\nu} = 8\pi T_{\mu\nu}, \tag{2.1}$$

where $G_{\mu\nu}$ is the Einstein tensor, $R_{\mu\nu} := R^{\lambda}_{\mu\lambda\nu}$ is the Ricci tensor derived from the Riemann curvature tensor, $R = g^{\mu\nu}R_{\mu\nu}$ is the Ricci scalar, $g_{\mu\nu}$ is the metric, $T_{\mu\nu}$ is the stress-energy tensor, and we assume natural units such that $G = c = 1$.

We consider the linearized treatment of GWs that Einstein originally studied (11; 12). In the following, Greek indices refer to 4-D spacetime coordinates, while Roman indices refer to 3-D spatial coordinates. The derivation presented here closely follows the treatment in Maggiore (29). We start with spatially-flat 4-D Minkowski spacetime, with $\eta_{\mu\nu} = \text{diag}(-1, 1, 1, 1)$, which upon adding a perturbation results in

$$g_{\mu\nu} = \eta_{\mu\nu} + h_{\mu\nu}, \tag{2.2}$$

where $h_{\mu\nu}$ is a tensor metric perturbation quantity that we treat only in the weak field case, $|h_{\mu\nu}| \ll 1$. This metric perturbation is assumed to be so small that the usual index raising and lowering operations can be performed with the Minkowski metric. Expanding the Einstein tensor to linear order in $h_{\mu\nu}$ in this perturbed spacetime gives,

$$G_{\mu\nu} = \frac{1}{2}\left(\partial_\mu\partial^\alpha h_{\alpha\nu} + \partial_\nu\partial^\alpha h_{\alpha\mu} - \partial_\mu\partial_\nu h - \Box h_{\mu\nu} + \eta_{\mu\nu}\Box h - \eta_{\mu\nu}\partial^\alpha\partial^\beta h_{\alpha\beta}\right), \tag{2.3}$$

where $\partial_\mu \equiv \partial/\partial x^\mu$; $\partial^\mu \equiv \eta^{\mu\nu}\partial_\nu$; $h = \eta^{\mu\nu}h_{\mu\nu}$ is the trace of $h_{\mu\nu}$; and $\Box = \eta^{\mu\nu}\partial_\mu\partial_\nu$ is the flat space d'Alembertian operator. This is a bit cumbersome, so we change variables in an effort to tidy this up. We define the trace-reversed perturbation, $\bar{h}_{\mu\nu} = h_{\mu\nu} - n_{\mu\nu}h/2$, which also implies that $\bar{h} = \eta^{\mu\nu}\bar{h}_{\mu\nu} = h - 2h = -h$, such that $h_{\mu\nu} = \bar{h}_{\mu\nu} - n_{\mu\nu}\bar{h}/2$. The Einstein tensor then becomes

$$G_{\mu\nu} = \frac{1}{2}\left(\partial_\mu\partial^\alpha \bar{h}_{\alpha\nu} + \partial_\nu\partial^\alpha \bar{h}_{\alpha\mu} - \Box\bar{h}_{\mu\nu} - \eta_{\mu\nu}\partial^\alpha\partial^\beta \bar{h}_{\alpha\beta}\right). \tag{2.4}$$

Similar to electromagnetism, we can ditch spurious degrees of freedom from our equations by choosing an appropriate gauge. If we consider the coordinate transformation $x^\alpha \mapsto x^\alpha + \xi^\alpha$, the transformation of the metric perturbation to first order is $h_{\mu\nu} \mapsto h_{\mu\nu} - (\partial_\mu\xi_\nu + \partial_\nu\xi_\mu)$. Asserting $\partial_\mu\xi_\nu$ to be of the same order as $|h_{\mu\nu}|$, the transformed metric perturbation retains the condition $|h_{\alpha\beta}| \ll 1$. This symmetry allows us to choose the *Lorenz gauge* (sometimes referred to as the *De Donder gauge*, or *Hilbert gauge*, or *harmonic gauge*) where $\partial^\nu\bar{h}_{\mu\nu} = 0$, such that the Einstein tensor reduces to the much more compact

$$G_{\mu\nu} = -\frac{1}{2}\Box\bar{h}_{\mu\nu}. \tag{2.5}$$

Note that in choosing the Lorenz gauge we have imposed 4 conditions that reduce the number of degrees of freedom in the symmetric 4×4 $h_{\mu\nu}$ matrix from 10 down to 6. Finally, the field equations reduce to

$$\Box \bar{h}_{\mu\nu} = -16\pi T_{\mu\nu}, \tag{2.6}$$

which can be generally solved using the radiative Green's function:

$$\bar{h}_{\mu\nu}(\vec{x}, t) = 4 \int d^4 x' \frac{T_{\mu\nu}(\vec{x}', t)}{|\vec{x} - \vec{x}'|}. \tag{2.7}$$

Examining Eq. 2.6 far from the source gives $\Box \bar{h}_{\mu\nu} = 0$, whose solution is clearly wave-like with propagation speed c. We might think that all components of the metric perturbation are radiative. This is a gauge artefact (13), where, in general, we can split the metric perturbation into (i) gauge degrees of freedom, (ii) physical, radiative degrees of freedom, and (iii) physical, non-radiative degrees of freedom. The Lorenz gauge is preserved under a coordinate transformation $x^\mu \mapsto x^\mu + \xi^\mu$, provided that $\Box \xi_\mu = 0$, which also implies that $\Box \xi_{\mu\nu} \equiv \Box(\partial_\mu \xi_\nu + \partial_\nu \xi_\mu - \eta_{\mu\nu} \partial_\rho \xi_\rho) = 0$. This means that the 6 independent components of $\bar{h}_{\mu\nu}$ can have $\xi_{\mu\nu}$ subtracted, which depend on 4 independent functions ξ_μ, thereby distilling the GW information down to 2 independent components of the metric perturbation. The conditions imposed on $\bar{h}_{\mu\nu}$ are such that $\bar{h} = 0$ (therefore $h_{\mu\nu} = \bar{h}_{\mu\nu}$), and $h^{0i} = 0$, which lead to the following properties of the metric perturbation that define the *transverse-traceless* (TT) gauge:

$$h^{0\mu} = 0, \quad h^i_i = 0, \quad \partial^j h_{ij} = 0. \tag{2.8}$$

Choosing a coordinate system such that we have a plane GW propagating in the z-direction in a vacuum, we can write the solution to $\Box \bar{h}_{\mu\nu} = 0$ as

$$h^{\mathrm{TT}}_{\mu\nu}(t, z) = \begin{pmatrix} 0 & 0 & 0 & 0 \\ 0 & h_+ & h_\times & 0 \\ 0 & h_\times & -h_+ & 0 \\ 0 & 0 & 0 & 0 \end{pmatrix} \cos\left[\omega(t - z)\right], \tag{2.9}$$

where h_+ and $h_\times$ are the amplitudes of the two distinct polarizations of GWs permitted within general relativity, denoted as "plus" (+) and "cross" (×) modes for how they tidally deform a circular ring of test masses in the plane perpendicular to the direction of propagation.

The tidal deformation of spacetime caused by a GW as it propagates is at the heart of all ground-based (LIGO-Virgo-KAGRA), Galactic-scale (pulsar-timing arrays), and planned space-borne (LISA) detection efforts. Consider the *coordinate* separation of two spatially-separated test masses as a GW sweeps by. The test masses fall along geodesics of the perturbed spacetime, such that in the weak-field regime, and to first order in amplitude of the wave, the coordinate separation of the test masses remains unchanged. However, the

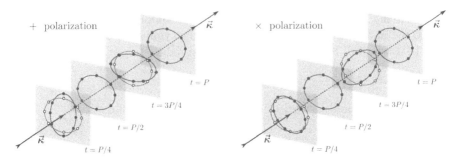

Figure 2.2: The periodic deformation of spacetime caused by a GW is illustrated for the two distinct polarizations permitted within GR. The influence of the GW is entirely in the plane perpendicular to the direction of propagation. Figure reproduced with permission from Ref. (30).

proper separation between the test masses *is* affected, and depends on the wave properties. The fractional change in the proper distance between two test masses separated by $\Delta x = L$ on the x-axis of a coordinate system due to the passage of a GW is given by $\delta L/L \simeq h_{xx}/2$, leading to a definition of the GW amplitude as the *strain*. For a periodic signal such that $h_{xx}(t, z = 0) = h_+ \cos \omega t$, we see that this proper distance separation oscillates according to $\delta \ddot{L} = -2\pi^2 f^2 L h_+ \cos \omega t$. We can see from Eq. 2.9 that the $+$-polarisation will lengthen distances along the x-axis while simultaneously contracting distances along the y-axis. The influence of the $h_\times$ mode is similar, but rotated by $\pi/4$ degrees counter-clockwise in the xy-plane. This tidal deformation in the plane perpendicular to the direction of propagation is illustrated for a full wave-cycle in Fig. 2.2.

2.2.2.1 *The Quadrupole Formula*

In the weak-field ($v \ll 1$) limit, far from a source, the leading order contribution to the solution of Eq. 2.6 is a function of the accelerating quadrupole moment of the source's mass distribution. Stress-energy conservation implies that the monopole makes no contribution due to conservation of the system's total energy, while the dipole moment makes no contribution due to conservation of momentum of the system's center-of-mass. Therefore the *quadrupole radiation formula* for the spatial components of the metric perturbation is (12),

$$h_{ij} = \frac{2}{r} \Lambda_{ij,kl} \frac{d^2 Q_{kl}}{dt^2}, \tag{2.10}$$

where r is the distance to the source, and Q_{kl} is the *reduced* quadrupole moment of $\rho(t, \vec{x})$ (the source's mass density), which is defined as,

$$Q_{ij} = \int d^3x \, \rho(t, \vec{x}) \left(x_i x_j - \frac{1}{3} r^2 \delta_{ij} \right), \tag{2.11}$$

and $\Lambda_{ik,jl}$ is the *Lambda tensor* that projects the metric perturbation in the Lorenz gauge into the TT gauge (see Ref. (29)).

The most relevant source system for us is a binary composed of compact objects (COs), whether white dwarfs, neutron stars, or black holes. Let us consider that each CO (of mass M) orbits one another at a distance $R(t)$ from their common center of mass (assumed to be far enough apart that we can ignore any tidal disruption effects) with slowly varying angular velocity $\omega(t)$. The COs are assumed to be moving at non-relativistic speeds to simplify the calculation of the quadrupole moment tensor. Kepler's third law gives us $\omega^2 = M/4R^3$, such that the total energy of the system is $E = -M^2/4R$. The orbital geometry is defined such that it lies in the $x - y$ plane with the origin coinciding with the system's center of mass. Each CO's coordinates at $t = 0$ are thus

$$x_1 = -x_2 = R\cos\omega t, \quad y_1 = -y_2 = R\sin\omega t, \quad z_1 = z_2 = 0. \qquad (2.12)$$

Evaluating the second time-derivative of the reduced quadrupole mass-moment, and plugging into Eq. 2.10, the radiative components of the GWs from this system are

$$h_{\mu\nu} = -\frac{8MR^2\omega^2}{r}\begin{pmatrix} 0 & 0 & 0 & 0 \\ 0 & \cos 2\omega t & \sin 2\omega t & 0 \\ 0 & \sin 2\omega t & -\cos 2\omega t & 0 \\ 0 & 0 & 0 & 0 \end{pmatrix}, \qquad (2.13)$$

where the GW frequency is *twice* the binary orbital frequency due to the quadrupolar nature of the emission. Note that the distance to the source is directly encoded in the amplitude of this leading-order radiative term.

Just like electromagnetic waves, or even waves on a string, the energy density in GWs is $\propto \omega^2 h^2$, where h is the amplitude of the wave. The amplitude is already proportional to a second time derivative of the reduced quadrupole moment, where the the factor of ω implies an additional time derivative. Therefore, based on simple scaling arguments, the energy density in GWs should be proportional to a quadratic combination of the third time-derivative of the quadrupole mass moment. In a background spacetime that is approximately flat far from the source, the Issacson stress-energy tensor of a GW is

$$T_{\mu\nu} = \frac{1}{32\pi}\langle \partial_\mu \hat{h}_{\alpha\beta} \partial_\nu \hat{h}^{\alpha\beta} \rangle, \qquad (2.14)$$

where $\langle \cdot \rangle$ denotes an average over several wave cycles. Evaluating the energy-flux elements of Eq. 2.14 in the quadrupole approximation, and integrating over the sphere, the GW luminosity of the source system

$$L_{\mathrm{GW}} \equiv \frac{dE_{\mathrm{GW}}}{dt} = \frac{1}{5}\left\langle \frac{d^3 Q_{ij}}{dt^3}\frac{d^3 Q^{ij}}{dt^3} \right\rangle, \qquad (2.15)$$

which, for a CO binary gives

$$L_{\rm GW} = \frac{128}{5}M^2R^4\omega^6 = \frac{128}{5}4^{1/3}\left(\frac{\pi M}{P}\right)^{10/3} \tag{2.16}$$

By equating the loss in the binary's orbital energy with GW emission, we can get a qualitative understanding of the frequency and strain evolution as the binary inspirals toward an eventual merger. The rate of change of the binary's orbital period is $dP/dt \propto P^{-5/3}$, such that the GW frequency evolution is $df_{\rm GW}/dt \propto f_{\rm GW}^{11/3}$. The strain amplitude is $h \propto R^2 f_{\rm GW}^2 \propto f_{\rm GW}^{2/3}$. Hence the orbital evolution, as driven by GW emission, causes the strain amplitude to increase in tune with the frequency, with both said to be "chirping" as the binary progresses toward merger.

2.3 STOCHASTIC GRAVITATIONAL WAVE BACKGROUNDS

Imagine yourself at a crowded party. All the guests are mingling and chatting to create a background hum of the usual social small-talk. You perk up your ears to attempt to hear whether your friend is stuck in an awkward conversation across the room, but alas, all you can hear is that damned background hum. Nevertheless, if we brought the lens of statistical inference to this party banter, there would be interesting information – the distribution of laughter could tell us whether this is a party worth sticking around at, and by localizing regions of high laughter one may be able to zero-in on the life and soul of this get-together. Occasionally, someone who may have had a bit too much to drink may forget to regulate their volume, creating a distinct sentence that can be heard above the fray.

Now imagine translating this scenario to GW signals. A cosmological population of systems emitting GWs of a similar frequency and comparable amplitude may not be able to be individually resolved by a detector. In such a scenario, the signals sum incoherently to produce a stochastic background of GWs (SGWB). "Stochastic" here formally means that we treat this background as a random process that is only studiable in terms of its statistical properties. As a function of time, the SGWB may look like random fluctuations without any discernible information, but if we dig deeper and look at its spectral information then it begins to reveal its secrets. It may have more power at lower frequencies than at higher frequencies, indicating that it varies on longer timescales. It may have a sharp uptick at a few frequencies, potentially revealing a few of those loud "voices" at the party. Let us now see how we describe a SGWB, and how it manifests as signals in our detectors. Much of the material in the following subsections has been adapted from the excellent treatment in Romano & Cornish (31).

2.3.1 The Energy Density of a SGWB

The energy content of the Universe today is dominated by three main components: dark energy makes up ~68%, dark matter makes up ~27%, while baryonic matter (everything we encounter in everyday life) and everything else makes up $\lesssim 5\%$. The fractional energy density is usually defined in terms of closure density, $\rho_c = 3H_0^2/8\pi$, which is the density of the Universe today that is required for flat spatial geometry, and where H_0 is Hubble's constant. For a SGWB, it makes most sense to consider the fractional energy density as a spectrum, in order to see how this energy density is distributed over frequencies that may or may not be accessible to our detectors. Hence, the energy-density spectrum in GWs is defined such that,

$$\Omega_{\text{SGWB}}(f) \equiv \frac{1}{\rho_c}\frac{d\rho}{d\ln f}, \tag{2.17}$$

where ρ is the energy density in GWs. The gauge invariant energy density is given by evaluating the 00 component of the stress-energy tensor, such that

$$T^{00} = \rho = \frac{1}{32\pi}\langle \dot{h}_{ab}\dot{h}^{ab}\rangle, \tag{2.18}$$

where an over-dot denotes ∂_t, and Roman indices denote the spatial components of the metric perturbation. This metric perturbation for a SGWB can be written in terms of an expansion over plane waves in the TT gauge,

$$h_{ab}(t,\vec{x}) = \sum_{A=+,\times}\int_{-\infty}^{\infty} df \int_{S^2} d^2\Omega_{\hat{n}}\; h_A(f,\hat{n})e^{2\pi i f(t+\hat{n}\cdot\vec{x})}e_{ab}^A(\hat{n}), \tag{2.19}$$

where $h_A(f,\hat{n})$ are complex random fields whose moments define the statistical properties of the stochastic GW background[3]; $\hat{n}$ is a unit vector pointing to the origin of the GW; and $e_{ab}^A(\hat{n})$ are the GW polarisation basis tensors, defined in terms of orthonormal basis vectors around $\hat{n}$:

$$e_{ab}^+(\hat{n}) = \hat{l}_a\hat{l}_b - \hat{m}_a\hat{m}_b, \quad e_{ab}^\times(\hat{n}) = \hat{l}_a\hat{m}_b + \hat{m}_a\hat{l}_b, \tag{2.20}$$

where

$$\hat{n} = \sin\theta\cos\phi\,\hat{x} + \sin\theta\sin\phi\,\hat{y} + \cos\theta\,\hat{z} \equiv \hat{r}$$
$$\hat{l} = \cos\theta\cos\phi\,\hat{x} + \cos\theta\sin\phi\,\hat{y} - \sin\theta\,\hat{z} \equiv \hat{\theta}$$
$$\hat{m} = -\sin\phi\,\hat{x} + \cos\phi\,\hat{y} \equiv \hat{\phi}. \tag{2.21}$$

The GW energy density is thus

$$\langle \dot{h}_{ab}\dot{h}^{ab}\rangle = \sum_A\sum_{A'}\int_{-\infty}^{\infty} df \int_{-\infty}^{\infty} df' \int_{S^2} d^2\Omega_{\hat{n}} \int_{S^2} d^2\Omega_{\hat{n}'}\; \langle h_A(f,\hat{n})h_{A'}^*(f',\hat{n}')\rangle$$
$$\times e_{ab}^A(\hat{n})e_{A'}^{ab}(\hat{n}') \times 4\pi^2 ff'$$
$$\times \exp\left[2\pi i(f-f')t + 2\pi i(\hat{n}-\hat{n}')\cdot\vec{x}\right]. \tag{2.22}$$

---

[3] Since $h_{ab}(t,\vec{x})$ is real, the Fourier amplitudes satisfy $h_a^*(f,\hat{n}) = h_A(-f,\hat{n})$

For a *Gaussian-stationary, unpolarized, spatially homogeneous and isotropic* stochastic background the quadratic expectation value of the Fourier modes are

$$\langle h_A(f,\hat{n})h_{A'}^*(f',\hat{n}')\rangle = \frac{\delta_{AA'}}{2}\frac{\delta^{(2)}(\hat{n},\hat{n}')}{4\pi}\frac{\delta(f-f')}{2}S_h(f), \qquad (2.23)$$

where $S_h(f)$ is the one-sided power spectral density (PSD) of the *Fourier modes* of the SGWB. With identities $\int d\hat{\Omega} = 4\pi$ and $\sum_A e_{ab}^A e_A^{ab} = 4$, converting the frequency integration bounds in Eq. 2.22 to $[0,\infty]$ gives

$$\langle \dot{h}_{ab}\dot{h}^{ab}\rangle = 8\pi^2 \int_0^\infty df\ f^2 S_h(f). \qquad (2.24)$$

Hence

$$\Omega_{\text{SGWB}}(f) \equiv \frac{1}{\rho_c}\frac{d\rho}{d\ln f} = \frac{2\pi^2}{3H_0^2}f^3 S_h(f). \qquad (2.25)$$

2.3.2 Characteristic Strain

The fractional energy density is often referenced in the cosmological literature, or even amongst particle physicists. But GW scientists typically talk about the *characteristic strain* of the SGWB, defined as

$$h_c(f) \equiv \sqrt{f S_h(f)}, \qquad (2.26)$$

The characteristic strain accounts for the number of wave cycles the signal spends in-band through the $\sqrt{f}$ dependence (see also Ref. (32)). Hence we can write

$$\Omega_{\text{SGWB}}(f) = \frac{2\pi^2}{3H_0^2}f^2 h_c^2(f). \qquad (2.27)$$

As is often the case in astronomy and astrophysics, several SGWB sources predict a power-law form for $h_c(f)$, defined as

$$h_c(f) = A_{\alpha,\text{ref}}\left(\frac{f}{f_{\text{ref}}}\right)^\alpha, \qquad (2.28)$$

where α is a spectral index, f_{ref} is a reference frequency that is typical of the detector's band, and $A_{\alpha,\text{ref}}$ is the characteristic strain amplitude at the reference frequency. The fractional energy density then scales as $\Omega_{\text{SGWB}}(f) \propto f^{2\alpha+2}$. Primordial SGWBs resulting from quantum tensor fluctuations that are inflated to macroscopic scales usually assume a scale-invariant spectrum for which $\Omega_{\text{SGWB}}(f) \propto$ constant, thereby implying $\alpha = -1$ for the characteristic strain spectrum. We'll see in Chapter 4 how a population of circular inspiraling compact-binary systems creates a SGWB with $\alpha = -2/3$ such that $h_c(f) \propto f^{-2/3}$ and $\Omega_{\text{SGWB}}(f) \propto f^{2/3}$.

2.3.3 Spectrum of the Strain Signal

We don't directly measure a metric perturbation in our detector; we measure the response of our detector to the influence of a metric perturbation. The strain signal measured by a detector is related to the source strain through the detector's response. The data in a given detector, $s_i(t)$, is a combination of measured signal, $h_i(t)$, and noise, $n_i(t)$:

$$s_i(t) = h_i(t) + n_i(t), \tag{2.29}$$

where the measured strain signal is the projection of the GW metric perturbation onto the detector's response, such that

$$h_i(t) = d_i^{ab} h_{ab}(\vec{x}_i, t), \tag{2.30}$$

and d_i^{ab} is the GW detector tensor of the ith detector. The one-sided cross-power spectral density (PSD) of the measured strain signal $S_s(f)$ in detectors i and j is

$$\langle \tilde{h}_i(f) \tilde{h}_j(f') \rangle = \frac{1}{2} \delta(f - f') S_s(f)_{ij}, \tag{2.31}$$

where tilde denotes a Fourier transform with the following convention:

$$\tilde{h}_i(f) = \int_{-\infty}^{\infty} e^{-2\pi i f t} h_i(t) dt. \tag{2.32}$$

The Fourier domain strain signal is thus

$$\tilde{h}_i(f) = \sum_A \int_{S^2} d^2\Omega_{\hat{n}} \int_{-\infty}^{\infty} df' \int_{-\infty}^{\infty} dt \, e^{-2\pi i f t} F_i^A(\hat{n}) h_A(f, \hat{n}) e^{2\pi i f'(t + \hat{n} \cdot \vec{x})}$$

$$= \sum_A \int_{S^2} d^2\Omega_{\hat{n}} \, F_i^A(\hat{n}) h_A(f, \hat{n}) e^{2\pi i f \hat{n} \cdot \vec{x}_i}, \tag{2.33}$$

where $F_i^A(\hat{n})$ is the antenna response beam pattern (or antenna pattern, or antenna response function) of the detector to mode-A of the GW, defined as

$$F_i^A(\hat{n}) \equiv d_i^{ab} e_{ab}^A(\hat{n}). \tag{2.34}$$

Hence, the quadratic expectation of the Fourier-domain signals is given by

$$\langle \tilde{h}_i(f) \tilde{h}_j^*(f') \rangle = \sum_A \sum_{A'} \int_{S^2} d^2\Omega_{\hat{n}} \int_{S^2} d^2\Omega_{\hat{n}'} \langle h_A(f, \hat{n}) h_{A'}^*(f', \hat{n}') \rangle$$

$$\times F_i^A(\hat{n}) F_j^{A'}(\hat{n}') e^{2\pi i (f \hat{n} \cdot \vec{x}_i - f' \hat{n}' \cdot \vec{x}_j)}, \tag{2.35}$$

which, upon using Eq. 2.23, becomes

$$\langle \tilde{h}_i(f) \tilde{h}_j^*(f') \rangle = \frac{1}{16\pi} S_h(f) \delta(f - f') \sum_A \int_{S^2} d^2\Omega_{\hat{n}} F_i^A(\hat{n}) F_j^{A'}(\hat{n}) e^{2\pi i f \hat{n} \cdot (\vec{x}_i - \vec{x}_j)}.$$

$$\tag{2.36}$$

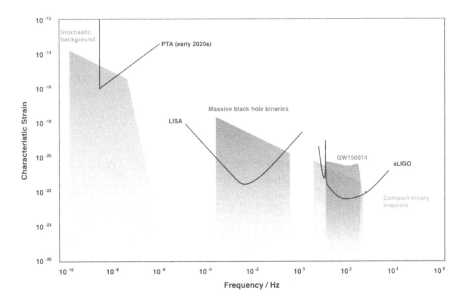

Figure 2.3: Overview of the detectors and sources covering the GW spectrum, spanning approximately twelve orders of magnitude in frequency. Figure created using http://gwplotter.com (32).

2.3.4 Overlap Reduction Function

If we compare Eq. 2.36 to Eq. 2.31, we see that

$$S_s(f)_{ij} = S_h(f) \times \frac{1}{8\pi} \sum_A \int_{S^2} d^2\Omega_{\hat{n}} F_i^A(\hat{n}) F_j^A(\hat{n}) e^{2\pi i f \hat{n} \cdot (\vec{x}_i - \vec{x}_j)}$$

$$= S_h(f) \times \tilde{\Gamma}_{ij}(f, \vec{x}_i - \vec{x}_j), \tag{2.37}$$

which relates the cross-power spectral density of the *measured strain signal* to that of the *Fourier modes of the GW background*. The term $\tilde{\Gamma}_{ij}(f, \vec{x}_i - \vec{x}_j)$ is referred to as the un-normalized overlap reduction function, or sometimes just the overlap reduction function (ORF) since we normalize it later as standard in order to be unity for co-located and oriented detectors. Essentially the ORF measures how much GWB power is shared between pairs of detectors. It is the most essential element of searching for a GWB with PTAs, and in the next chapter we will see what this PTA ORF is (spoiler: it's the ubiquitous *Hellings & Downs curve!* (33)).

2.4 THE GRAVITATIONAL WAVE SPECTRUM

Like electromagnetic radiation, GWs come in a spectrum of frequencies from many different sources, where (roughly speaking) the characteristic frequency of the emitting system scales inversely with its total mass. However, electro-magnetic radiation from astronomical sources is usually an incoherent super-position of emission from regions much larger than the radiation wavelength. By contrast, the majority of the GW spectrum being targeted by current and forthcoming detectors is sub-kHz, and as such wavelengths are of comparable scale to their emitting systems. Thus GWs directly track the coherent bulk motions of relativistic compact objects.

This volume is not intended to be a comprehensive overview of the enor-mous efforts being exerted towards detection across the full GW landscape. Nevertheless, there are three main detection schemes blanketing the spec-trum from nanoHertz frequencies (with Pulsar Timing Arrays, the focus of this volume), milliHertz frequencies (with laser interferometry between drag-free space-borne satellites), to 10–100 s Hz (ground-based laser interferome-try). At higher frequencies there are efforts to detect MegaHertz GWs with smaller instruments (e.g., 34), while even lower than Pulsar Timing Arrays at $\sim 10^{-15}$ Hz there are ongoing efforts to detect GWs through the imprint of curl-mode patterns in the polarization signal of the Cosmic Microwave Back-ground (35; 36; 28).

Figure 2.3 illustrates the strain spectrum of GWs over many decades in frequency, with the sensitivity of current and planned detectors overlaid on the bands of compelling astrophysical targets. At the lowest frequencies we need precisely timed pulsars at $\sim$kiloparsec distances to probe nHz frequen-cies, where a stochastic background of merging supermassive binary black holes may be found. The $\sim 1 - 10^4$ Hz band is the domain of terrestrial de-tectors, where kilometer-scale laser interferometers target the chirping signals of inspiraling stellar-mass compact binary systems, as well as core-collapse supernovae. The lowest frequency we can probe with terrestrial detectors (~ 1 Hz) is restricted by local gravitational gradients and seismic noise iso-lation. The only way to overcome this "seismic wall" is by moving to space. Space-borne laser interferometers of $\sim 10^9$ m arm-length are planned which will dig into the $0.1 - 100$ mHz band, allowing for massive black hole mergers to be seen throughout the entire Universe, precision tests of gravity, and the entire Galactic white-dwarf binary population to be seen.

Below is a brief review of ground-based and space-borne efforts. The Galactic-scale efforts of Pulsar Timing Arrays will be discussed in the re-maining chapters of this volume.

2.4.1 Ground-based Detectors

Early terrestrial detectors were of the Weber bar variety (37; 38), however it was soon realised that laser interferometry had the potential to greatly surpass the relatively narrow-band sensitivity of bars. The Weber bar was principally developed to infer the *transfer of energy* of GWs to the detection apparatus, while laser interferometery aims to measure the strain influence of a GW on the propagation of a precisely monitored laser beam. If we consider a GW impinging on a simple right-angled interferometer, with laser beams propagating along orthogonal directions, then the wave's influence is measured via the alternate stretching and compression of the proper-length of the arms, inducing a phase-shift in the recombined laser beams. As such, contemporary GW detectors aim to measure the *strain amplitude* of a passing wave.

GW interferometers aim to achieve a strain sensitivity of $\sim 10^{-21}$ or lower. The basic mode of operation is that of a Michelson interferometer, where laser light is injected into the interferometer and subsequently split into two beams to propagate along orthogonal arms. Each laser beam reflects off end test-mass optics and recombine at a beamsplitter to interfere at a photodiode. There are many sources of noise (see, e.g., 39, and references therein), including photon shot noise and thermal motion of the reflecting test masses. The limiting noise sources at the lowest frequencies are seismic and gravity-gradient noise. The former can be ameliorated by a combination of pendulum isolation, spring suspension, and anti-vibration actuators. However gravity-gradient noise is caused by seismic waves that create local mass density fluctuations, whose gravitational influence couples to the test-masses; this can be minimised by monitoring seismic activity to subtract its signal, or moving the detector underground, but below ~ 1 Hz the detector must be completely distanced from these surface-wave density fluctuations by moving to space.

LIGO (the Laser Interferometer Gravitational wave Observatory)[4] is the first kilometre-scale detection instrument for GWs, and is operated in partnership between the California Institute of Technology and the Massachusetts Institute of Technology. There are two 4 km arm-length instruments in total, all located in the USA, with one sited in Hanford, Washington,[5] and one in Livingston, Louisiana. In addition to LIGO, there is the French/Italian 3 km Virgo interferometer[6] located at Cascina, near Pisa, in Italy. Also in Europe is the 600 m arm-length GEO-600 interferometer[7], located near Hannover, Germany. With a smaller baseline and lower laser power, GEO-600 cannot match the sensitivity of LIGO/Virgo, but it has been a useful testbed for advanced technologies and techniques. Finally, in Japan there is the 3 km arm-length KAmioka GRAvitational wave telescope (KAGRA)[8], which operates in the Kamioka mine under cryogenic conditions. Placing KAGRA underground dramatically suppresses seismic disturbances and gravity gradient noise.

Beyond these second-generation detector plans, there are concepts for new third-generation detectors aiming to achieve a broadband order of magnitude improvement in strain sensitivity and to push operation down into the

---

[4] https://www.ligo.caltech.edu

[5] In the pre-2015 Initial and Enhanced LIGO stages there was a second half-length Hanford instrument contained within the same vacuum envelope

[6] https://www.virgo-gw.eu

[7] https://www.geo600.org

[8] https://gwcenter.icrr.u-tokyo.ac.jp/en/

$\sim 1 - 10$ Hz range[9]. The most notable of these are the Einstein Telescope in Europe (40)[10], and Cosmic Explorer in the USA (41)[11].

Sources – In its most recent incarnation, LIGO became an officially funded NSF project in 1994 under the leadership of Barry Barish. Construction broke ground in Hanford, Washington in 1994, and Livingston, Louisiana in 1995, with initial observations beginning in 2002. For approximately eight years of these initial observations LIGO did not see a single shred of evidence for GWs. However, this was not totally unexpected – Initial LIGO was a dress rehearsal that acted as a technology concept and a rallying site for the community, but a remote prospect for a first detection. Around 2010, LIGO was taken offline and subjected to a multi-year overhaul to create Advanced LIGO, which saw first operations commence in 2015.

The rest is now popular lore. On September 14, 2015, before the first science run officially commenced, LIGO detected the collision of two $\sim 30 M_\odot$ black holes at a distance of $\sim$400 Mpc from Earth, resulting in the emission of $\sim 3 M_\odot$ of mass-energy as GWs, leaving a remnant black hole of $\sim 57 M_\odot$ (42). The signal lasted $\sim$200 milliseconds in the sensitivity band of the detector, providing a signal-to-noise ratio of $\sim$24. Far from being the threshold event that had been expected for the first signal, this arrived with thunderous certainty. GW150914 (labeled by its discovery date) was announced to the world on February 11, 2016, inaugurating the field of direct GW astronomy. Since then, many other BH-BH signals have been detected, including GW190521 that has the largest constituent masses to date of $\sim 85 M_\odot$ and $\sim 66 M_\odot$ (43). These BH-BH detections have an enormous amount to teach us about their possible origin pathways (44; 45; 46), including as isolated binary stellar systems that undergo successive supernovae and a common stellar-envelope phase, or as systems formed through dynamical capture in a stellar cluster environment, or even scenarios we can't envision yet. What's more, on August 17, 2017, LIGO and Virgo observed a double neutron-star collision, yielding not only a GW signal registered at all three sites, but a plethora of electromagnetic signals observed across the spectrum (47). GW170817 provided extraordinary multi-messenger insight into neutron star astrophysics (48), tests of General Relativity (49), and even an anchor in spacetime with which we can calibrate the expansion rate of the Universe (50).

It's not just compact-binary coalescences that ground-based detectors can find. Other searches include modelled and unmodeled bursts (from core-collapse supernovae) (e.g., 51; 52), continuous waves (from <10-cm pulsar "mountains" that generate a quadrupole mass moment) (e.g., 53; 54; 55), and an unresolved background of GWs (e.g., 56; 57). As of writing, LIGO is offline due to the COVID-19 pandemic.

---

[9]https://gwic.ligo.org/3Gsubcomm/documents/science-case.pdf
[10]http://www.et-gw.eu
[11]https://cosmicexplorer.org

2.4.2 Space-borne Detectors

Access to the frequency range $\sim 0.1 - 100$ mHz requires much longer interferometer arm lengths, and a complete suppression of the seismic and gravity-gradient noise that plagues the low-frequency operation of terrestrial detectors. To this end, detection at these frequencies necessitates space-borne laser interferometers.

The canonical design for a mission in this band is the Laser Interferometer Space Antenna (LISA)[12,13], a project led by the European Space Agency in collaboration with the National Aeronautics and Space Administration. This mission calls for an arrangement of three identical satellites maintaining a triangular constellation, each separated by 2.5×10^9 m. The test masses within each satellite are expected to be ~ 46 mm, 2 kg, gold-coated cubes of gold/platinum. Each satellite would achieve zero drag, where each test mass essentially floats in free fall while the surrounding satellite absorbs local gravitational influences. Micro-thrusters reposition the satellite around the test mass to maintain this drag-free configuration. The LISA satellite constellation would trail the Earth's motion by $20°$ as it orbits around the Sun. With two optical links in each arm of the triangle, a total of six optical links should allow for the interferometer to operate in Sagnac mode, constructing a data-stream that will be completely insensitive to laser, optical-bench, and clock noise (e.g., 58). In 2013, the European Space Agency selected "The Gravitational Universe" science theme for its L3 mission slot[14], wherein it committed to launch a space-borne GW mission, due for launch in 2034 at the earliest. In 2017, LISA was proposed and accepted as the candidate mission[15].

In 2015, a mission known as LISA Pathfinder (LPF)[16] was launched to act as a technology demonstration for the LISA drag-free concept. Rather than three satellites, LPF consisted of just one satellite enclosing an optical system that corresponded to an arm-length of 38-cm. It reached its designated position at Lagrange point L1 on January 22, 2016. Far from a mere technology demonstration, LPF exceeded its scientific mandate by achieving exceptional noise precision that is comparable to the LISA requirement (59).

Despite LISA's prospective launch date being (as of writing) more than a decade away, possible follow-up missions are already being studied and advocated for. These missions are designed to bridge either the $\sim 1 - 100$ μHz gap between PTAs and LISA (60), or the $0.1 - 10$ Hz gap between LISA and ground-based detectors (61). Some notable examples include μAres (60), a straw-person μHz detector concept suggested in the European Space Agency's *Voyage 2050* long-range planning call. This ambitious mission would place at

---

[12] https://www.elisascience.org

[13] https://lisa.nasa.gov

[14] http://www.esa.int/Science_Exploration/Space_Science/ESA_s_new_vision_to_study_the_invisible_Universe

[15] https://www.lisamission.org/?q=news/top-news/gravitational-wave-mission-selected-esas-l3-mission

[16] https://sci.esa.int/web/lisa-pathfinder

least three satellites in an approximately Martian orbit with the Sun at its center. At the deciHertz level, the Advanced Laser Interferometer Antenna (ALIA) (62; 63), and DECi-hertz Interferometer Gravitational wave Observatory (DECIGO) mission (64) have been proposed to capitalize on the treasure trove of science related to stellar-mass binary black holes, intermediate mass-ratio inspirals, and binary neutron stars that would otherwise be missed.

Sources – This frequency regime is a rich astrophysical zoo of sources, including a collection of Galactic white-dwarf binary systems whose properties are well-known electromagnetically, and hence should be detectable within a few weeks or months of instrument operation (65). As a sure bet for GW detection by LISA, these systems are often referred to as *verification binaries*, in that if no signal is observed then something is very wrong with the instrument. Additionally, ~ 25000 other ultra-compact white-dwarf binary systems may be individually resolvable in GWs (66; 67), while the remaining several million will create a stochastic GW foreground signal (e.g., 68; 69).

Massive BH binary systems in the mass range $10^4 - 10^7 M_\odot$ are prime targets for LISA, which should detect the inspiral, merger, and ringdown signals associated with coalescence (70). Detailed parameter estimation of individual systems will be possible, permitting inference of such effects like spin-orbit precession, higher waveform harmonics, and eccentricity (e.g., 71; 72; 73; 74). The detection rate is highly variable, ranging from ~ 0.5 per year to ~ 100 per year depending on the underlying "seed" formation scenario and growth factors (e.g., 75; 76). Nevertheless a catalog of such systems will empower demographic and "genealogical" studies of the massive BH population, unveiling the factors that drive their growth over cosmic time, and the fingerprints of their initial formation at high redshift (e.g., 77; 78).

At lower masses, a tantalizing possibility for LISA is that it will capture the very early inspiral stage of stellar-mass ($\mathcal{M} \gtrsim 25 M_\odot$) and intermediate-mass ($10^2 M_\odot < M < 10^4 M_\odot$) BH-BH systems that will eventually also register (after $\sim$months) at higher frequencies in future ground-based detectors. By detecting the same systems at low frequencies (wide separation) and high frequencies (close separation through merger), multi-band GW astrophysics will yield insights into the dynamics of stellar-mass BH-BH over long timescales (see, e.g., 79, and references therein). Combining the information from both detection schemes will break important degeneracies in parameter inference, allowing the systems to be characterized better (80; 81). Early work in this area emphasized how LISA detection of these systems would allow precise time and sky-location forecasts of the eventual merger in order for ground-based detectors to be on the watch. But the reverse scheme is also important: detection of these systems in a future ground-based detector will allow archival LISA detector to be mined for marginal or sub-threshold signals.

From a fundamental physics standpoint, arguably the most exciting sources in the LISA band are the Extreme Mass-Ratio Inspirals, where stellar-mass compact remnants gradually spiral in towards a much larger BH, and in so doing map out the geometry of the massive BH's space-time (see, e.g., 82, and references therein). Beyond these extreme mass-ratio systems, LISA has huge potential to probe how modifications to General Relativity affect GW generation, GW propagation, BH spacetimes, and BH dynamics (83). Indeed, even if no hints of GR departures are found, LISA may detect more exotic signals than mere compact binaries; a cosmological GW background signal of primordial origin may be detectable at these frequencies (e.g., 84), or even a similar background formed later as the Universe undergoes phase transitions (e.g., 85). LISA could even potentially detect cosmic strings that form during these phase transitions; these strings can intersect one another to chop off small loops, which then emit GWs as they vibrate relativistically under extreme tension (86).

As of writing, LISA's prospective launch date in 2034 is 13 years away. But this does not diminish the vigor and excitement with which the GW and astrophysics communities are currently exploring its enormous science potential.

Bibliography

[1] I Newton, A Motte, and NW Chittenden. *Newton's Principia: The Mathematical Principles of Natural Philosophy.* Nineteenth Century Collections Online (NCCO): Science, Technology, and Medicine: 1780–1925. Geo. P. Putnam, 1850. 2.1.1

[2] Clifford M Will. The confrontation between general relativity and experiment. *Living reviews in relativity*, 9(1):3, 2006. 2.1.2

[3] Robert V Pound and Glen A Rebka Jr. Apparent weight of photons. *Physical Review Letters*, 4(7):337, 1960. 2.1.2

[4] Irwin I Shapiro. Fourth test of general relativity. *Physical Review Letters*, 13(26):789, 1964. 2.1.2

[5] Bruno Bertotti, Luciano Iess, and Paolo Tortora. A test of general relativity using radio links with the cassini spacecraft. *Nature*, 425(6956):374–376, 2003. 2.1.2

[6] CW Francis Everitt, DB DeBra, BW Parkinson, et al. Gravity probe b: final results of a space experiment to test general relativity. *Physical Review Letters*, 106(22):221101, 2011. 2.1.2

[7] James G Williams, Slava G Turyshev, and Dale H Boggs. Progress in lunar laser ranging tests of relativistic gravity. *Physical Review Letters*, 93(26):261101, 2004. 2.1.2

[8] SM Ransom, IH Stairs, AM Archibald, et al. A millisecond pulsar in a stellar triple system. *Nature*, 505(7484):520–524, January 2014. 2.1.2

[9] Anne M Archibald, Nina V Gusinskaia, Jason WT Hessels, et al. Universality of free fall from the orbital motion of a pulsar in a stellar triple system. *Nature*, 559(7712):73–76, July 2018. 2.1.2

[10] Henri Poincaré. *Sur la dynamique de l'électron*. Circolo Matematico di Palermo, 1906. 2.2.1

[11] Albert Einstein. Approximative integration of the field equations of gravitation. *Sitzungsber. Preuss. Akad. Wiss. Berlin (Math. Phys.)*, 1916(688–696):1, 1916. 2.2.1, 2.2.2

[12] Albert Einstein. Über gravitationswellen. *SPAW*, pages 154–167, 1918. 2.2.1, 2.2.2, 2.2.2.1

[13] Arthur Stanley Eddington. The propagation of gravitational waves. *Proceedings of the Royal Society of London. Series A, Containing Papers of a Mathematical and Physical Character*, 102(716):268–282, 1922. 2.2.1, 2.2.2

[14] Albert Einstein. Einstein versus the physical review. *Physics Today*, 58(9):43, 2005. 2.2.1

[15] Albert Einstein and Nathan Rosen. On gravitational waves. *Journal of the Franklin Institute*, 223(1):43–54, 1937. 2.2.1

[16] Felix AE Pirani. On the physical significance of the riemann tensor. *AcPP*, 15:389–405, 1956. 2.2.1

[17] Hermann Bondi. Plane gravitational waves in general relativity. *Nature*, 179(4569):1072–1073, 1957. 2.2.1

[18] Joseph Weber. Evidence for discovery of gravitational radiation. *Physical Review Letters*, 22(24):1320, 1969. 2.2.1

[19] Joseph Weber. Anisotropy and polarization in the gravitational-radiation experiments. *Physical Review Letters*, 25(3):180, 1970. 2.2.1

[20] RA Hulse and JH Taylor. Discovery of a pulsar in a binary system. *The Astrophysical Journal*, 195:L51–L53, January 1975. 2.2.1

[21] JH Taylor and JM Weisberg. A new test of general relativity - Gravitational radiation and the binary pulsar PSR 1913+16. *The Astrophysical Journal*, 253:908–920, February 1982. 2.2.1

[22] M Kramer and IH Stairs. The double pulsar. *Annual Review of Astronomy and Astrophysics*, 46:541–572, September 2008. 2.2.1

[23] John Antoniadis, Paulo CC Freire, Norbert Wex, et al. A Massive Pulsar in a Compact Relativistic Binary. *Science*, 340(6131):448, April 2013. 2.2.1

[24] JM Weisberg and Y Huang. Relativistic Measurements from Timing the Binary Pulsar PSR B1913+16. *The Astrophysical Journal*, 829(1):55, September 2016. 2.1

[25] GE Moss, LR Miller, and RL Forward. Photon-noise-limited laser transducer for gravitational antenna. *Applied Optics*, 10(11):2495–2498, 1971. 2.2.1

[26] Rainer Weiss and Dirk Muehlner. Electronically coupled broadband gravitational antenna. *Quarterly Progress Report, Research Laboratory of Electronics*, 105, 1972. 2.2.1

[27] BICEP2 Collaboration, PAR Ade, RW Aikin, et al. Detection of B-Mode Polarization at Degree Angular Scales by BICEP2. *Physical Review Letters*, 112(24):241101, June 2014. 2.2.1

[28] BICEP2 Collaboration, Keck Array Collaboration, PAR Ade, et al. Constraints on Primordial Gravitational Waves Using Planck, WMAP, and New BICEP2/Keck Observations through the 2015 Season. *Physical Review Letters*, 121(22):221301, November 2018. 2.2.1, 2.4

[29] Michele Maggiore. *Gravitational waves: Volume 1: Theory and experiments*, volume 1. Oxford university press, 2008. 2.2.2, 2.2.2.1

[30] Nigel T Bishop and Luciano Rezzolla. Extraction of gravitational waves in numerical relativity. *Living Reviews in Relativity*, 19(1):2, October 2016. 2.2

[31] Joseph D Romano and Neil J Cornish. Detection methods for stochastic gravitational-wave backgrounds: a unified treatment. *Living Reviews in Relativity*, 20(1):2, April 2017. 2.3

[32] CJ Moore, RH Cole, and CPL Berry. Gravitational-wave sensitivity curves. *Classical and Quantum Gravity*, 32(1):015014, January 2015. 2.3.2, 2.3

[33] RW Hellings and GS Downs. Upper limits on the isotropic gravitational radiation background from pulsar timing analysis. *The Astrophysical Journal*, 265:L39–L42, February 1983. 2.3.4

[34] Aaron S Chou, Richard Gustafson, Craig Hogan, et al. MHz gravitational wave constraints with decameter Michelson interferometers. *Physical Review D*, 95(6):063002, March 2017. 2.4

[35] Marc Kamionkowski, Arthur Kosowsky, and Albert Stebbins. A Probe of Primordial Gravity Waves and Vorticity. *Physical Review Letters*, 78(11):2058–2061, March 1997. 2.4

[36] Uroš Seljak and Matias Zaldarriaga. Signature of Gravity Waves in the Polarization of the Microwave Background. *Physical Review Letters*, 78(11):2054–2057, March 1997. 2.4

[37] Joseph Weber and John A Wheeler. Reality of the cylindrical gravitational waves of einstein and rosen. *Reviews of Modern Physics*, 29(3):509, 1957. 2.4.1

[38] Joseph Weber. Detection and generation of gravitational waves. *Physical Review*, 117(1):306, 1960. 2.4.1

[39] BP Abbott, R Abbott, TD Abbott, et al. A guide to LIGO-Virgo detector noise and extraction of transient gravitational-wave signals. *Classical and Quantum Gravity*, 37(5):055002, March 2020. 2.4.1

[40] Michele Maggiore, Chris Van Den Broeck, Nicola Bartolo, et al. Science case for the Einstein telescope. *Journal of Cosmology and Astroparticle Physics*, 2020(3):050, March 2020. 2.4.1

[41] David Reitze, Rana X Adhikari, Stefan Ballmer, et al. Cosmic Explorer: The U.S. Contribution to Gravitational-Wave Astronomy beyond LIGO. In *Bulletin of the American Astronomical Society*, volume 51, page 35, September 2019. 2.4.1

[42] BP Abbott, R Abbott, TD Abbott, et al. Observation of Gravitational Waves from a Binary Black Hole Merger. *Physical Review Letters*, 116(6):061102, February 2016. 2.4.1

[43] R Abbott, TD Abbott, S Abraham, et al. GW190521: A Binary Black Hole Merger with a Total Mass of 150 $M_\odot$. *Physical Review Letters*, 125(10):101102, September 2020. 2.4.1

[44] BP Abbott, R Abbott, TD Abbott, et al. Astrophysical implications of the binary black-hole merger GW150914. *The Astrophysical Journal*, 818:L22, February 2016. 2.4.1

[45] R Abbott, TD Abbott, S Abraham, et al. Population properties of compact objects from the second LIGO–Virgo gravitational-wave transient catalog. *The Astrophysical Journal Letters*, 913(1):L7, 2021. 2.4.1

[46] BP Abbott, R Abbott, TD Abbott, et al. Binary black hole population properties inferred from the first and second observing runs of advanced LIGO and advanced virgo. *The Astrophysical Journal*, 882(2):L24, September 2019. 2.4.1

[47] BP Abbott, R Abbott, TD Abbott, et al. GW170817: Observation of gravitational waves from a binary neutron star inspiral. *Physical Review Letters*, 119(16):161101, October 2017. 2.4.1

[48] BP Abbott, R Abbott, TD Abbott, et al. Multi-messenger observations of a binary neutron star merger. *The Astrophysical Journal*, 848(2):L12, October 2017. 2.4.1

[49] BP Abbott, R Abbott, TD Abbott, et al. Tests of general relativity with GW170817. *Physical Review Letters*, 123(1):011102, July 2019. 2.4.1

[50] BP Abbott, R Abbott, TD Abbott, et al. A gravitational-wave standard siren measurement of the Hubble constant. *Nature*, 551(7678):85–88, November 2017. 2.4.1

[51] BP Abbott, R Abbott, TD Abbott, et al. First targeted search for gravitational-wave bursts from core-collapse supernovae in data of first-generation laser interferometer detectors. *Physical Review D*, 94(10):102001, November 2016. 2.4.1

[52] The LIGO Scientific Collaboration, the Virgo Collaboration, BP Abbott, et al. All-sky search for short gravitational-wave bursts in the second Advanced LIGO and Advanced Virgo run. *arXiv e-prints*, page arXiv:1905.03457, May 2019. 2.4.1

[53] BP Abbott, R Abbott, TD Abbott, et al. First narrow-band search for continuous gravitational waves from known pulsars in advanced detector data. *Physical Review D*, 96(12):122006, December 2017. 2.4.1

[54] BP Abbott, R Abbott, TD Abbott, et al. Searches for continuous gravitational waves from 15 Supernova Remnants and Fomalhaut b with advanced LIGO. *The Astrophysical Journal*, 875(2):122, April 2019. 2.4.1

[55] BP Abbott, R Abbott, TD Abbott, et al. All-sky search for continuous gravitational waves from isolated neutron stars using Advanced LIGO O2 data. *Physical Review D*, 100(2):024004, July 2019. 2.4.1

[56] BP Abbott, R Abbott, TD Abbott, et al. Upper limits on the stochastic gravitational-wave background from advanced LIGO's first observing run. *Physical Review Letters*, 118(12):121101, March 2017. 2.4.1

[57] BP Abbott, R Abbott, T D Abbott, et al. Search for the isotropic stochastic background using data from Advanced LIGO's second observing run. *Physical Review D*, 100(6):061101, September 2019. 2.4.1

[58] Daniel A Shaddock. Operating LISA as a Sagnac interferometer. *Physical Review D*, 69(2):022001, January 2004. 2.4.2

[59] Michele Armano, Heather Audley, Gerard Auger, et al. Sub-femto-g free fall for space-based gravitational wave observatories: Lisa pathfinder results. *Physical review letters*, 116(23):231101, 2016. 2.4.2

[60] Alberto Sesana, Natalia Korsakova, Manuel Arca Sedda, et al. Unveiling the Gravitational Universe at micro-Hz Frequencies. *arXiv e-prints*, page arXiv:1908.11391, August 2019. 2.4.2

[61] Manuel Arca Sedda, Christopher PL Berry, Karan Jani, et al. The missing link in gravitational-wave astronomy: discoveries waiting in the decihertz range. *Classical and Quantum Gravity*, 37(21):215011, November 2020. 2.4.2

[62] Peter L Bender, Mitchell C Begelman, and Jonathan R Gair. Possible LISA follow-on mission scientific objectives. *Classical and Quantum Gravity*, 30(16):165017, August 2013. 2.4.2

[63] Guido Mueller, John Baker, Simon Barke, et al. Space based gravitational wave astronomy beyond LISA. In *Bulletin of the American Astronomical Society*, volume 51, page 243, September 2019. 2.4.2

[64] Shuichi Sato, Seiji Kawamura, Masaki Ando, et al. The status of DE-CIGO. In *Journal of Physics Conference Series*, volume 840 of *Journal of Physics Conference Series*, page 012010, May 2017. 2.4.2

[65] T Kupfer, V Korol, S Shah, et al. LISA verification binaries with updated distances from Gaia Data Release 2. *Monthly Notices of the Royal Astronomical Society*, 480(1):302–309, October 2018. 2.4.2

[66] Tyson B Littenberg, Katelyn Breivik, Warren R Brown, et al. Astro2020 decadal science white paper: Gravitational wave survey of galactic ultra compact binaries. *arXiv e-prints*, page arXiv:1903.05583, March 2019. 2.4.2

[67] Pau Amaro-Seoane, Heather Audley, Stanislav Babak, et al. Laser Interferometer Space Antenna. *arXiv e-prints*, page arXiv:1702.00786, February 2017. 2.4.2

[68] V Korol, S Toonen, A Klein, et al. Populations of double white dwarfs in Milky Way satellites and their detectability with LISA. *Astronomy & Astrophysics*, 638:A153, June 2020. 2.4.2

[69] Travis Robson and Neil Cornish. Impact of galactic foreground characterization on a global analysis for the LISA gravitational wave observatory. *Classical and Quantum Gravity*, 34(24):244002, December 2017. 2.4.2

[70] Monica Colpi, Kelly Holley-Bockelmann, Tamara Bogdanovic, et al. Astro2020 science white paper: The gravitational wave view of massive black holes. *arXiv e-prints*, page arXiv:1903.06867, March 2019. 2.4.2

[71] Joey Shapiro Key and Neil J Cornish. Characterizing spinning black hole binaries in eccentric orbits with LISA. *Physical Review D*, 83(8):083001, April 2011. 2.4.2

[72] Antoine Klein, Yannick Boetzel, Achamveedu Gopakumar, et al. Fourier domain gravitational waveforms for precessing eccentric binaries. *Physical Review D*, 98(10):104043, November 2018. 2.4.2

[73] Neil J Cornish and Kevin Shuman. Black hole hunting with LISA. *Physical Review D*, 101(12):124008, June 2020. 2.4.2

[74] Sylvain Marsat, John G Baker, and Tito Dal Canton. Exploring the Bayesian parameter estimation of binary black holes with LISA. *arXiv e-prints*, page arXiv:2003.00357, February 2020. 2.4.2

[75] Michael L Katz, Luke Zoltan Kelley, Fani Dosopoulou, et al. Probing massive black hole binary populations with LISA. *Monthly Notices of the Royal Astronomical Society*, 491(2):2301–2317, January 2020. 2.4.2

[76] Priyamvada Natarajan, Angelo Ricarte, Vivienne Baldassare, et al. Disentangling nature from nurture: tracing the origin of seed black holes. *Bulletin of the American Astronomical Society*, 51(3):73, May 2019. 2.4.2

[77] Alberto Sesana, Jonathan Gair, Emanuele Berti, and Marta Volonteri. Reconstructing the massive black hole cosmic history through gravitational waves. *Physical Review D*, 83(4):044036, February 2011. 2.4.2

[78] Joseph E Plowman, Ronald W Hellings, and Sachiko Tsuruta. Constraining the black hole mass spectrum with gravitational wave observations - II. Direct comparison of detailed models. *Monthly Notices of the Royal Astronomical Society*, 415(1):333–352, July 2011. 2.4.2

[79] Curt Cutler, Emanuele Berti, Kelly Holley-Bockelmann, et al. What can we learn from multi-band observations of black hole binaries? *Bulletin of the American Astronomical Society*, 51(3):109, May 2019. 2.4.2

[80] Atsushi Nishizawa, Emanuele Berti, Antoine Klein, and Alberto Sesana. eLISA eccentricity measurements as tracers of binary black hole formation. *Physical Review D*, 94(6):064020, September 2016. 2.4.2

[81] Salvatore Vitale. Multiband Gravitational-Wave Astronomy: Parameter Estimation and Tests of General Relativity with Space- and Ground-Based Detectors. *Physical Review Letters*, 117(5):051102, July 2016. 2.4.2

[82] Christopher Berry, Scott Hughes, Carlos Sopuerta, et al. The unique potential of extreme mass-ratio inspirals for gravitational-wave astronomy. *Bulletin of the American Astronomical Society*, 51(3):42, May 2019. 2.4.2

[83] Emanuele Berti, Enrico Barausse, Ilias Cholis, et al. Tests of General Relativity and Fundamental Physics with Space-based Gravitational Wave Detectors. *Bulletin of the American Astronomical Society*, 51(3):32, May 2019. 2.4.2

[84] Angelo Ricciardone. Primordial Gravitational Waves with LISA. In *Journal of Physics Conference Series*, volume 840 of *Journal of Physics Conference Series*, page 012030, May 2017. 2.4.2

[85] Chiara Caprini, Mikael Chala, Glauber C Dorsch, et al. Detecting gravitational waves from cosmological phase transitions with LISA: an update. *Journal of Cosmology and Astroparticle Physics*, 2020(3):024, March 2020. 2.4.2

[86] Pierre Auclair, Jose J Blanco-Pillado, Daniel G Figueroa, et al. Probing the gravitational wave background from cosmic strings with lisa. *Journal of Cosmology and Astroparticle Physics*, 2020(04):034, 2020. 2.4.2

CHAPTER 3

Pulsar Timing

3.1 PULSARS

Pulsars are extraordinary. They are a special class of neutron star, which in themselves are mind-boggling objects corresponding to the collapsed $\sim M_\odot$ cores of massive stars that have undergone supernovae, leaving only small $\sim 10–15$ km carcasses that are supported against total collapse by neutron degeneracy pressure. Since their discovery in 1967 by Jocelyn Bell Burnell, Antony Hewish and collaborators (1), pulsars have shed light on strong-field gravity, the equation of state of nuclear matter, evolutionary scenarios for massive binary systems, the structure of the ionized interstellar medium, the existence of exoplanets, and much more. It would be difficult to overstate the exquisite astrophysical laboratories presented to us in the form of isolated and relativistic-binary pulsars. For deeper reviews, see Refs. (2; 3; 4). Various other sources have inspired the content of this chapter, including Verbiest et al. (2021) (5) and Burke-Spolaor (2015) (6).

The "lighthouse model" provides our basic framework for understanding and modeling pulsars as rapidly rotating, highly magnetized neutron stars resulting from stellar collapse (7; 8). Due to conservation of angular momentum and magnetic flux, these pulsars are far more rapidly spinning and magnetized than their progenitor stars. Their magnetic field (whose axis may not necessarily align with its rotational axis) is such that the star acts as a rotating magnetic dipole, generating a local electric field along which charged particles within the co-rotating magnetospheric plasma are accelerated. It is expected that these particles excite beams of radio emission high in the pulsar atmosphere that we observe whenever the rotating beam intersects our line-of-sight (9; 10). The pulse period is then a measure of the rotation period of the pulsar itself.

While the largest population of pulsars are in the class of initially discovered young ~ 1-second rotators (so-called "canonical pulsars"), they are not used in precision timing campaigns for GW searches. The broad reasons for this are that canonical pulsars can exhibit lower long-timescale rotational stability, and glitches, the latter of which are a technical term for a sudden

DOI: 10.1201/9781003240648-3

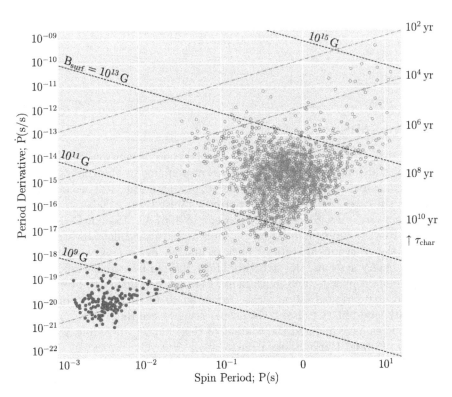

Figure 3.1: $P - \dot{P}$ diagram of known pulsars. MSPs are filled green circles, while canonical pulsars are open circles. Data were taken from the ATNF pulsar catalogue (11) version 1.56 (see also www.atnf.csiro.au/research/pulsar/psrcat). The dashed black lines show estimated surface magnetic fields strengths (B_{surf}), while the dot-dashed grey lines show the lines of characteristic age (τ_{char}). Figure reproduced with permission from Ref. (12).

spin-up in the pulsar that *may* be related to reconnection of the internal neutron superfluid with the crustal lattice (e.g., 13, and references therein). The 1982 discovery of the pulsar B1937+21, with its 1.5 millisecond period, was the first of the new class of "millisecond pulsars" (MSPs) (14). The demography of pulsars can be broadly split into these two varieties (see Fig. 3.1); the canonical pulsars are ones that have formed relatively recently as a result of a supernova, while the millisecond pulsars are older, having spun down and been subsequently recycled back to millisecond periods via the accretion of material and angular momentum during mass transfer from a binary companion (15).

3.2 PRECISION PULSAR TIMING

The key to using pulsars as astrophysical tools is that they can be used as excellent time-keepers[1]. We observe pulses of radio emission separated by the observational period of the pulsar. However, the shape of each pulse from one rotation to the next varies randomly, possibly associated with stochasticity in the emission region through which our line of sight is intersecting. But the pulse shape *averaged* over rotations is remarkably stable and reproducible on timescale from minutes to decades (17)[2]. It is this stability at a given radio frequency that permits precision timing; the pulse shape is unique to each pulsar and can be relied upon to mark the passage of rotations when receiving a train of radio pulses.

A schematic diagram of the main stages involved in pulsar timing is shown in Fig. 3.2. Upon being accelerated in the pulsar's magnetosphere, high-energy charged particles excite beams of radiation with a steep, negative-slope radio spectrum. This radiation propagates through the ionized interstellar medium (ISM), suffering dispersion and other radio-frequency dependent delays. Dispersion arises from the frequency-dependent refractive index of the ISM, such that lower radio frequencies have a reduced group velocity, arriving at the telescope later than higher radio-frequency components of the radiation. The delay is determined by the distance traveled through the ISM, such that with an appropriate model of the line-of-sight electron-density distribution, the measured dispersion can be used to infer the pulsar's distance (e.g., 18, and references therein). Dispersion can be overcome either by splitting the observed band into smaller sub-channels and delaying the higher-frequency components according to the dispersive $1/\nu^2$ relationship (incoherent dedispersion), or by convolving the raw observations with the inverse transfer function of the ISM (coherent dedispersion) (19). Further details of these effects, and how their residual influences are modeled alongside GWs, are given in Chapter 7.

After the removal of dispersion, thousands of pulses are integrated over ~minutes to an hour of observation, and folded by the current estimated

---

[1] Full details of timing procedures can be found in Ref. (16).

[2] Investigations of the evolution of the standard pulse profile can yield rich information on the details of the emission region, and geodetic precession of the pulsar's spin axis in a binary system. See Ref. (4).

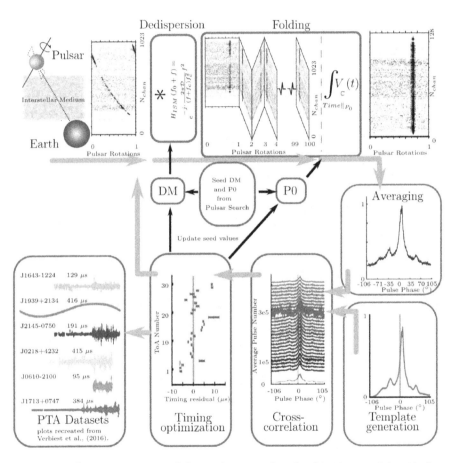

Figure 3.2: A representation of the main stages involved in the precision timing of pulsars. Figure reproduced with permission from Ref. (12).

rotational period to give a boost to the signal strength and stabilize the mea-sured pulse profile. This measured pulse is then cross correlated with the *template profile* for the specific pulsar at the specific observing frequency. The phase offset between the measured pulse and template profile is added to the time-stamp of the observation, giving the pulse *"Time Of Arrival"* (TOA). The template-fitting uncertainty of the phase offset, and thus the measure-ment uncertainty of the TOA, scales as

$$\sigma_{\text{TOA}} \propto \frac{1}{\sqrt{n_P t_{\text{int}} \Delta\nu}} \times \frac{T_{\text{sys}}}{T_{\text{peak}}} \times \frac{P}{\sqrt{W(P-W)}}, \tag{3.1}$$

where n_P is the number of combined polarizations, t_{int} is the integration time, $\Delta\nu$ is the radio bandwidth of observation, T_{peak} is the brightness temperature of the pulsar at the peak of its profile, T_{sys} is the brightness temperature (i.e., noise) of the observing system, W is the integrated pulse intensity divided by its peak intensity (the pulse's *equivalent width*), and P is the pulse period. The TOA measurement uncertainty is often referred to as "radiometer noise", corresponding to the limiting precision with which we can time a pulsar due to the radio telescope's sensitivity and the properties of the pulsar itself. The radiometer noise is not the end of the story for the per-TOA measurement uncertainty, as various modifying parameters are introduced to correct poorly-estimated σ_{TOA} values or account for additional sources of epoch noise; this is further described in Chapter 7.

Over many repeated observations, a train of pulses are collected and the TOA computed for each as described above. The next stage is determining the *timing model* for the pulsar (sometimes also referred to as the *timing ephemeris*). This is a generative model that describes all deterministic influ-ences that affect the arrival time of the pulses as they propagate from the pulsar system to the radio telescope on Earth. We write the pulse TOA as

$$t_{\text{PSR}} = t_{\text{TOA}} - \Delta_{\odot} - \Delta_{\text{IISM}} - \Delta_B, \tag{3.2}$$

where t_{PSR} is the pulse emission time at the pulsar and t_{TOA} is the pulse arrival time at the radio telescope. The term $\Delta_{\odot}$ accounts for timing correc-tions back to the quasi-inertial reference frame of the Solar System Barycenter (SSB) that include (a) Einstein delays due to time dilation and gravitational redshift in the presence of the Sun and other bodies in the Solar System; (b) Shapiro delays due to light propagating through the gravitational potential well of the Sun; (c) Roemer delays due to the classic light travel time across the Solar System from the Earth to the SSB; (d) Earth atmospheric propaga-tion delays; (e) solar-wind induced radio-frequency-dependent delays; and (f) clock corrections from the observatory standards to global timing standards. The term Δ_{IISM} includes corrections for radio-frequency dependent propaga-tion delays such as dispersion. While the pulse components are dispersion-corrected before the folding stage, the effects of interstellar dispersion have been shown to vary over long and short timescales, requiring additional mod-eling as described in detail in Chapter 7. Finally, for pulsars in binary systems, a further transformation, Δ_B, is needed to correct for time between the binary barycenter and the emitting pulsar itself. This includes Einstein, Shapiro, and Roemer delays within the pulsar binary orbit, in addition to a host of other higher-order corrections (see, e.g., 20, and references therein).

After all of these corrections, the final model of the pulsar's phase evolution is remarkably simple. The lighthouse model describes a beam sweeping into our line-of-sight every pulsar rotation, where the rotational frequency of the pulsar is decreasing due to "spindown" that may be related to the electromagnetic outflow tapping the pulsar's rotational kinetic energy. Hence, for a pulsar with some rotational frequency $1/P$ measured at epoch t_0, the pulse phase is modeled as

$$\phi(t_{\mathrm{PSR}}) = \phi_0 + 2\pi \frac{(t_{\mathrm{PSR}} - t_0)}{P} - \frac{1}{2} 2\pi \left[\frac{(t_{\mathrm{PSR}} - t_0)}{P} \right]^2 \dot{P} + \dots, \qquad (3.3)$$

where ϕ_0 is the pulsar phase at t_0. With initial estimates of the dispersion measure, rotational period, period derivative, and location of the pulsar, we can perform a least-squares fit of the collection of measured TOAs with the field-standard software package TEMPO2 (21; 20; 22) or the emerging heir-apparent PINT (23). The differences between the measured TOAs and the predictions of the best-fit model are called the *timing residuals*. By iterating and refining the timing model to remove systematic trends and miminize the residuals, we can construct extraordinarily precise predictions of the pulsar's phase.

By definition, the residuals are generated by any phenomena that are not included in the timing model. Ideally this would be only radiometer noise (and GWs of course!), but there are many other sources of noise and uncertainty in pulsar-timing observations. For example, some pulsars are known to exhibit rotational irregularities. As mentioned earlier, discrete jumps in the rotational frequency of the pulsar (glitches) are thought to occur as a result of the sudden recoupling and angular momentum transfer between the neutron-superfluid and the crustal lattice, reducing the lag in their rotational frequencies which occurs due to the minimal friction between the two (24). This glitchy behaviour is suppressed in older and millisecond pulsars (25; 26), so it is less of a concern for PTA GW searches. More relevant, however, is the fact that many pulsars exhibit timing noise with low-frequency structure (*red timing noise*). The origin of this may be due to the pulsar's magnetosphere rapidly and sporadically switching between stable configurations, leading to different pulse shapes and spindown rates (27). The variation in spindown rate causes the rotational frequency to wander over a period of years, contributing a source of red timing noise if unmodeled. While magnetospheric mode switching/nulling is not incorporated into the timing model, it can be accounted for as an extra red stochastic process (28). Likewise, time-varying electron densities along the line-of-sight to each pulsar can result in time-dependent dispersion measure that can manifest as a radio-frequency dependent red stochastic process in the timing residuals (e.g., 29; 30, and references therein). Further details of these and other noise sources, as well as the respective approaches we adopt to model them, are given in Chapter 7. Ultimately, the target of our PTA searches is an effect that is not included in the pulsar timing model, specifically because this target phenomena influences all monitored pulsars in a correlated fashion. This target is of course the timing deviations induced by GWs.

The following sections describe the timing response of a pulsar to the influence of a GW signal, the inter-pulsar correlation pattern that we hunt for in searches for a stochastic background of GWs, and also additional sources of timing errors that could produce an apparent inter-pulsar correlated signal.

3.3   TIMING RESPONSE TO GRAVITATIONAL WAVES

We exploit the precision timing of millisecond pulsars to directly search for GWs, treating the pulsar and the SSB respectively as opposite ends of our laboratory setup. A passing GW perturbs the spacetime metric along the Earth-pulsar line of sight (31; 32; 33; 34), deforming the proper separation, and thereby inducing irregularities in the perceived pulsar rotational frequency. We provide a derivation of the pulsar timing response due to a transiting GW below; this closely follows the treatment in Maggiore, Volume 2 (35). We use the following line element for our GW spacetime:

$$ds^2 = -dt^2 + [\delta_{ab} + h_{ab}^{\mathrm{TT}}(t, \vec{x})]dx^a dx^b. \tag{3.4}$$

where we follow the convention that Roman indices a, b denote spatial components of the metric, while i, j denote different pulsars. For a photon path traveling along the x-axis toward an observer at the origin, we have that $ds^2 = 0$, and thus

$$dx \approx -\left\{1 - \frac{1}{2} h_{xx}^{\mathrm{TT}}[t, \vec{x}(t)]\right\} dt. \tag{3.5}$$

Integrating both sides for an Earth-pulsar coordinate separation of L, we have

$$L = t_{\mathrm{obs}} - t_{\mathrm{em}} - \frac{1}{2} \int_{t_{\mathrm{em}}}^{t_{\mathrm{obs}}} dt' \, h_{xx}^{\mathrm{TT}}[t', \vec{x}(t')]. \tag{3.6}$$

Given that h_{xx}^{TT} is a small quantity, we are permitted to use $t_{\mathrm{obs}} \approx t_{\mathrm{em}} + L$ in the integral, with the photon path being approximately along its unperturbed trajectory $\vec{x}(t) = (t_{\mathrm{obs}} - t)\hat{p}$. We can also generalize to an arbitrary pulsar location by replacing h_{xx}^{TT} with $p^a p^b h_{ab}^{\mathrm{TT}}$. Thus

$$t_{\mathrm{obs}} = t_{\mathrm{em}} + L + \frac{p^a p^b}{2} \int_{t_{\mathrm{em}}}^{t_{\mathrm{em}}+L} dt' \, h_{ab}^{\mathrm{TT}}[t', (t_{\mathrm{em}} + L - t')\hat{p}]. \tag{3.7}$$

We now consider the arrival time of a subsequent pulse emitted after one rotational period of the pulsar, such that $t'_{\mathrm{em}} = t_{\mathrm{em}} + P$. The observed arrival time of this second pulse is simply given by substituting $t_{\text{em}} \mapsto t_{\text{em}} + L$ in Eq. 3.7, and subtracting to get

$$t'_{\text{obs}} - t_{\text{obs}} = P + \Delta P, \tag{3.8}$$

where

$$\Delta P = \frac{p^a p^b}{2} \int_{t_{\text{em}}}^{t_{\text{em}}+L} dt' \left\{ h_{ab}^{\text{TT}}[t' + P, \vec{x}_0(t')] - h_{ab}^{\text{TT}}[t', \vec{x}_0(t')] \right\}, \tag{3.9}$$

where $\vec{x}_0(t') = (t_{\text{em}} + L - t')\hat{p}$. The observed arrival time difference between two subsequent pulses is thus equal to the spin period of the pulsar, plus an extra GW-induced term. The spin period of the pulsars are $\sim$ milliseconds, whereas the GW periods of interest span months to decades. Hence, the first integrand term inside the curly brackets can be Taylor expanded to first order, leaving

$$\frac{\Delta P}{P} = \frac{p^a p^b}{2} \int_{t_{\text{em}}}^{t_{\text{em}}+L} dt' \left[\frac{\partial}{\partial t'} h_{ab}^{\text{TT}}(t', \vec{x}) \right]_{\vec{x}=\vec{x}_0(t')}. \tag{3.10}$$

Let us now consider a fiducial monochromatic wave solution propagating along the direction $\hat{\Omega}$:

$$h_{ab}^{\text{TT}}(t', \vec{x}) = \mathcal{A}_{ab}(\hat{\Omega}) \cos \left[\omega_{\text{GW}}(t' - \hat{\Omega} \cdot \vec{x}) \right]. \tag{3.11}$$

Upon substituting into Eq. 3.10, we get

$$\frac{\Delta P}{P} = \frac{1}{2} \frac{p^a p^b \mathcal{A}_{ab}}{(1 + \hat{\Omega} \cdot \hat{p})} \left\{ \cos \left[\omega_{\text{GW}} t_{\text{obs}} \right] - \cos \left[\omega_{\text{GW}} t_{\text{em}} - \omega_{\text{GW}}(t_{\text{obs}} - t_{\text{em}})\hat{\Omega} \cdot \hat{p} \right] \right\}$$

$$= \frac{1}{2} \frac{p^a p^b \mathcal{A}_{ab}}{(1 + \hat{\Omega} \cdot \hat{p})} \left\{ \cos \left[\omega_{\text{GW}} t_{\text{obs}} \right] - \cos \left[\omega_{\text{GW}}(t_{\text{em}} - L\hat{\Omega} \cdot \hat{p}) \right] \right\}. \tag{3.12}$$

We now define the GW-induced redshift of the pulse arrival rate as $z(t) \equiv (\nu_0 - \nu(t))/\nu_0 = -(\Delta \nu / \nu) = \Delta P / P$. From the previous equation, and with a GW propagating in direction $\hat{\Omega}$, this can be written as

$$z(t, \hat{\Omega}) = \frac{1}{2} \frac{p^a p^b}{(1 + \hat{\Omega} \cdot \hat{p})} \left[h_{ab}(t, \vec{x}_{\text{earth}}) - h_{ab}(t - L, \vec{x}_{\text{pulsar}}) \right]$$

$$= \frac{1}{2} \frac{p^a p^b}{(1 + \hat{\Omega} \cdot \hat{p})} \Delta h_{ab} \tag{3.13}$$

where $t = t_{\text{obs}}$, the position vector of the pulsar is $\vec{x}_{\text{pulsar}} = L\hat{p}$, and the position vector of the Earth (or more precisely, the Solar System Barycenter) is $\vec{x}_{\text{earth}} = 0$. Thus the GW imparts two redshifting signatures on the pulse arrival times: an imprint of the metric perturbation as the wave washes over the Earth (the so-called *Earth term*), and an imprint of the retarded metric perturbation from when the wave washed over the pulsar (the *pulsar term*).

The time difference between the pulsar term and Earth term is of order the light travel time of the Earth-pulsar distance, which for pulsars at ~ kiloparsec distances can be thousands of years. This presents a very exciting possibility, since the imprinted GW signature carries a measure of the emitting source from thousands of years in the past. When compounded over many timed pulsars, this essentially creates a form of temporal aperture synthesis to allow the source evolution to be tracked over baselines much larger than our timing campaigns.

Given Eq. 2.19, and assuming that the amplitude of the metric perturbation at the time of passing the pulsar and Earth is unchanged, we can write Δh_{ab} as

$$\Delta h_{ab} = \int_{-\infty}^{\infty} df \left[e^{2\pi i f t} \left(e^{-2\pi i f L(1+\hat{\Omega}\cdot\hat{p})} - 1 \right) \times \sum_A h_A(f,\hat{\Omega}) e_{ab}^A(\hat{\Omega}) \right], \quad (3.14)$$

which has as its Fourier transform

$$\Delta \tilde{h}_{ab}(f,\hat{\Omega}) = \left(e^{-2\pi i f L(1+\hat{\Omega}\cdot\hat{p})} - 1 \right) \sum_A h_A(f,\hat{\Omega}) e_{ab}^A(\hat{\Omega}). \quad (3.15)$$

We will later be interested in the Fourier transform of $z(t)$ in Eq. 3.13, which is thus

$$\tilde{z}(f,\hat{\Omega}) = \left(e^{-2\pi i f L(1+\hat{\Omega}\cdot\hat{p})} - 1 \right) \sum_A h_A(f,\hat{\Omega}) F^A(\hat{\Omega}), \quad (3.16)$$

where $F^A(\hat{\Omega})$ is the Ath-mode GW antenna response pattern (see Eq. 2.34) for an Earth-pulsar system, defined as

$$F^A(\hat{\Omega}) = \frac{1}{2} \frac{p^a p^b}{(1+\hat{\Omega}\cdot\hat{p})} e_{ab}^A(\hat{\Omega}). \quad (3.17)$$

The GW-induced timing perturbations measured with respect to a reference time $t = 0$ are defined as

$$R(t) \equiv \int_0^t dt' \, z(t') = \frac{1}{2} \frac{p^a p^b}{(1+\hat{\Omega}\cdot\hat{p})} \int_0^t dt' \, [h_{ab}(t, \vec{x}_{\text{earth}}) - h_{ab}(t - L, \vec{x}_{\text{pulsar}})]$$

$$= R(t)_{\text{earth}} - R(t)_{\text{pulsar}}. \quad (3.18)$$

3.4 OVERLAP REDUCTION FUNCTION FOR A BACKGROUND OF GRAVITATIONAL WAVES

PTAs target a stochastic background of gravitational waves as the expected primary signal class. More details of why this is will be given in the next chapter, but suffice it to say that the GWs we are sensitive to are very low frequency, weakly evolving, and our detector's frequency resolution is also fairly limited. If the GW signal is composed of discrete systems, then their respective signals will likely pile up within each frequency resolution bin, defying our ability to full resolve them. Hence, in a practical search scenario, we need to understand how much this GWB will induce correlated power in the timing deviations between pulsars that are widely separated across the sky. We thus need to compute the PTA overlap reduction function (ORF) for a GWB. We draw from Section 3.3 and Section 2.3.4 to compute this. The derivation has been given several times in the literature (e.g., 36; 37; 38; 39), but we align most closely with that given in Maggiore, Volume 2 (35). An important caveat is that this derivation is for an *isotropic* GWB, with unifrom distribution of power across the sky. Further discussions of anisotropy and modifications for alternative GW polarizations are given toward the end of this section.

We recall that Eq. 3.16 gives the GW-induced redshift of the pulse arrival rate in the Fourier domain. If we take the integral of this over the entire sky to account for GWs coming from any direction, and compute the expectation of the inter-pulsar correlation of these redshifts over many random SGWB realizations, we get

$$\langle \tilde{z}_i(f)\tilde{z}_j^*(f') \rangle = \int_{S^2} \int_{S'^2} d^2\hat{\Omega}\, d^2\hat{\Omega}' \left[e^{-2\pi i f L_i(1+\hat{\Omega}\cdot\hat{p}_i)} - 1 \right] \left[e^{2\pi i f' L_j(1+\hat{\Omega}'\cdot\hat{p}_j)} - 1 \right]$$
$$\times \left\langle \sum_A h_A(f,\hat{\Omega}) F^A(\hat{\Omega}) \sum_{A'} h_{A'}^*(f',\hat{\Omega}') F^{A'}(\hat{\Omega}') \right\rangle. \quad (3.19)$$

We can now invoke the assumption of a Gaussian, stationary, isotropic, and unpolarized SGWB using Eq. 2.23, which reduces the above equation to

$$\langle \tilde{z}_i(f)\tilde{z}_j^*(f') \rangle = \frac{1}{2}\delta(f-f')S_s(f)_{ij}$$
$$= \frac{1}{2}\delta(f-f')S_h(f)\int_{S^2}\frac{d^2\hat{\Omega}}{8\pi}\,\kappa_{ij}(f,\hat{\Omega})\sum_{A=+,\times}F_i^A(\hat{\Omega})F_j^A(\hat{\Omega}),$$
$$(3.20)$$

where

$$\kappa_{ij}(f,\hat{\Omega}) = \left[e^{-2\pi i f L_i(1+\hat{\Omega}\cdot\hat{p}_i)} - 1 \right]\left[e^{2\pi i f L_j(1+\hat{\Omega}\cdot\hat{p}_j)} - 1 \right]. \quad (3.21)$$

Thus the cross-power spectral density of the measured redshift to the pulse arrival rate is

$$S_s(f)_{ij} = \frac{1}{2}S_h(f)\int_{S^2}\frac{d^2\hat{\Omega}}{4\pi}\,\kappa_{ij}(f,\hat{\Omega})\sum_{A=+,\times}F_i^A(\hat{\Omega})F_j^A(\hat{\Omega}), \quad (3.22)$$

with units of [time]. It's worth discussing $\kappa_{ij}(f,\hat{\Omega})$, which controls how rapidly the pulsar term of the measured signal spatially decorrelates. Even the closest known millisecond pulsars are >100 parsecs distant from us, while the minimum GW frequency that we can probe with current IPTA datasets is $\sim 10^{-9}$ Hz, which still leaves $fL > 10$. The complex exponentials are thus rapidly oscillating terms that contribute negligibly to the integral above, except in the case where the pulsars are *identical* (i.e., same distance and sky location). Hence, $\kappa_{ij}(f, \hat{\Omega}) \to 2$ when $i = j$, and $\kappa_{ij}(f, \hat{\Omega}) \to 1$ otherwise. See Mingarelli & Sidery (2014) (40) for a complete discussion of this aspect. Bearing this in mind for later, we consider $i \neq j$ in the following.

Comparing Eq. 3.22 with Eq. 2.37, we see that our goal is to calculate the un-normalized ORF, $\tilde{\Gamma}_{ij}(f)$, for PTAs, which under our approximation for $\kappa_{ij}(f, \hat{\Omega})$, is frequency independent. While elementary, any which way you do this integral is going to be tedious. I recommend following the detailed derivations cited in the initial paragraph above. The result is

$$\tilde{\Gamma}_{ij} = \int_{S^2} \frac{d^2\Omega_{\hat{n}}}{4\pi} \sum_{A=+,\times} F_i^A(\hat{n}) F_j^A(\hat{n})$$

$$= x_{ij} \ln(x_{ij}) - \frac{1}{6} x_{ij} + \frac{1}{3}, \tag{3.23}$$

where $x_{ij} = (1 - \cos(\theta_{ij}))/2$ and θ_{ij} is the angular separation between the position of pulsars on the sky. This was the initial calculation and normalization presented by Hellings & Downs (41) (although it was initially presented in that paper without derivation steps). In the PTA literature, you'll more often see the general expression that accounts for $i = j$, and normalizes the ORF such that $\Gamma_{ij} = 1$ for $i = j$. This expression is

$$\Gamma_{ij} = \frac{3}{2} x_{ij} \ln(x_{ij}) - \frac{1}{4} x_{ij} + \frac{1}{2} + \frac{1}{2} \delta_{ij}, \tag{3.24}$$

where δ_{ij} is the Kronecker delta function. This function is shown in Fig. 3.3 along with some notable features. We get maximal values in the pulsar autocorrelations. Even pulsars with very small angular separations will have a cross-correlation no greater than 0.5, since the pulsar terms decorrelate rapidly once they are spatially separated by approximately more than a GW wavelength. The Hellings & Downs curve exhibits a strong quadrupolar trend over angular separation as a result of the quadrupolar GW antenna response patterns of the Earth-pulsar systems. However, it is not a pure quadrupole; the fact that the curve only returns to 0.25 at 180° instead of 0.5 is evidence of that. This is because the denominator in the antenna response pattern (see Eq. 3.17) introduces a preferred direction, where the response is largest to GWs propagating parallel to the radio pulses traveling from the pulsar to Earth. In fact, we can perform a decomposition of the Hellings & Downs curve in terms of Legendre polynomials to inspect the power in different multipoles. The result is (38; 42; 43)

$$\Gamma_{ij} = \sum_{l=0}^{\infty} a_l P_l(\cos \theta_{ij}), \tag{3.25}$$

where $i \neq j$, $a_0 = 0 = a_1$, and

$$a_l = \frac{3}{2} \frac{(l-2)!}{(l+2)!} (2l + 1). \tag{3.26}$$

Implicit in this decomposition is that the distribution of pulsars is isotropic across the sky, giving a distribution of pulsar angular separations that is $\propto \sin\theta_{ij}$. The Legendre spectrum for the Hellings & Downs curve is shown in Fig. 3.4, with all power contained in $l \geq 2$, meaning that the Hellings & Downs curve is orthogonal to monopole and dipole inter-pulsar correlations in the limit of infinite precision data and uniform pulsar sky coverage. In fact, as expected, the quadrupole dominates such that $a_2/\sum_{l=0}^{\infty} a_l = 0.63$, and the octupole adds another 0.17; hence 80% of the Hellings & Downs curve can be described in terms of $l = 2$ and $l = 3$ Legendre polynomials.

The Hellings & Downs curve is not the end of the story for cross-correlation SGWB searches with PTAs. As mentioned, it is only the ORF for an *isotropic* SGWB, and also implicitly assumes GR as the correct theory of gravity through the transverse-tensor nature of the antenna response. Significant work has been done to generalize the ORF calculation to anisotropic SGWBs (37; 44; 45; 38), allowing the angular structure to be probed using techniques similar to the Cosmic Microwave Background (38; 46; 47), and for more bespoke methods to be developed that target discrete systems constituting the background signal (48; 49). An anisotropic ORF is no longer merely a function of the angular separation of pulsars on the sky, since the expected signal cross-correlation will also be dependent on the pulsar positions with respect to the distribution of angular GW power. Likewise, in a general metric theory of gravity, there are a total of six allowed GW polarization states: two transverse-tensor (TT) GR states (TT$_+$, TT$_\times$), one scalar transverse state (ST), one scalar longitudinal state (SL), and two vector longitudinal (VL) states (VL$_x$, VL$_y$). The presence of these alternative polarization states modifies the ORF, and in the cases of the longitudinal modes make the ORF unavoidably dependent on the pulsar distances and GW frequency (50; 51; 52; 53; 54). It is also possible to show that a massive graviton will distort the expected Hellings & Downs curve through its effect on the graviton dispersion relation (55; 56).

Given Eq. 3.24, the cross-power spectral density of the redshift to the pulse arrival rate can finally be written as

$$S_s(f)_{ij} = \frac{1}{3}\Gamma_{ij} S_h(f). \tag{3.27}$$

However, we don't measure shifted pulse arrival rates. We measure pulse arrival times, and their deviations away from modeled predictions. As shown in Eq. 3.18, the GW-induced timing perturbation is an accumulating shift over time. This will simply result in factors of $1/(2\pi i f)$ (and the conjugate) when the Δh_{ij} factor of Eq. 3.14 is integrated over time, and hence a factor of $1/4\pi^2 f^2$ when these timing perturbations are correlated. Thus the cross-power spectral density of the GW-induced timing deviations is

$$S_t(f)_{ij} = \frac{S_s(f)_{ij}}{4\pi^2 f^2} = \Gamma_{ij}\frac{S_h(f)}{12\pi^2 f^2} = \Gamma_{ij}\frac{h_c^2(f)}{12\pi^2 f^3}, \tag{3.28}$$

which has units of [time]^3, and where we have used the previously introduced

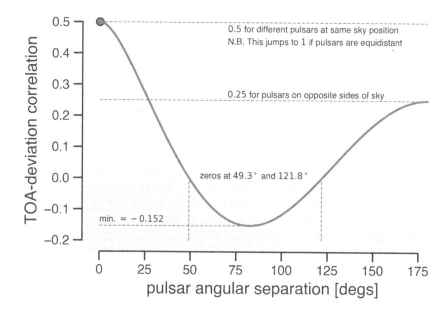

Figure 3.3: The normalized overlap reduction function for an isotropic SGWB in PTAs, more commonly referred to as *The Hellings & Downs Curve*, since it was first shown in Hellings & Downs (1983) (41). Some instructive features of the curve are labeled.

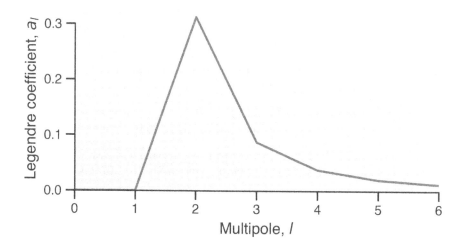

Figure 3.4: Legendre polynomial spectrum of the Hellings & Downs curve (38; 42; 43). 63% of the power is contained in $l = 2$, while 80% is contained within $l = 2$ and $l = 3$.

relationship $S_h(f) = h_c^2(f)/f$. We can convert this process from the Fourier domain into the time domain using the *Wiener-Khinchin Theorem*:

$$C(\tau)_{ij} = \int_0^\infty df \, S_t(f)_{ij} \cos(2\pi f \tau), \qquad (3.29)$$

which gives the covariance of GW-induced timing deviations between pulsars i and j over a lag time between observations, τ. This covariance has the expected units of $[\text{time}]^2$.

As a final note, the SGWB is not the only process that can induce low-frequency inter-pulsar correlated timing deviations in PTA data. We will see in the next chapter that systematic errors in our model of the Solar System ephemeris could induce dipolar correlated timing deviations through incorrect TOA barycentering. Likewise, long timescale drifts in global timing standards would affect all pulsars equally, inducing monopole correlated timing deviations. Fortunately, since the Hellings & Downs curve is orthogonal to these correlations, with sufficient pulsar sky coverage and data precision we can isolate our SGWB searches from these systematics.

Bibliography

[1] Antony Hewish, S Jocelyn Bell, John DH Pilkington, et al. Observation of a rapidly pulsating radio source. In *Pulsating Stars*, pages 5–9. Springer, 1968. 3.1

[2] DR Lorimer and M Kramer. *Handbook of Pulsar Astronomy*, volume 4. 2004. 3.1

[3] Duncan R Lorimer. Binary and millisecond pulsars. *Living Reviews in Relativity*, 11(1):8, November 2008. 3.1

[4] Ingrid H Stairs. Testing general relativity with pulsar timing. *Living Reviews in Relativity*, 6(1):5, September 2003. 3.1, 2

[5] JPW Verbiest, S Oslowski, and S Burke-Spolaor. Pulsar timing array experiments. *arXiv e-prints*, page arXiv:2101.10081, January 2021. 3.1

[6] Sarah Burke-Spolaor. Gravitational-wave detection and astrophysics with pulsar timing arrays. *arXiv e-prints*, page arXiv:1511.07869, November 2015. 3.1

[7] F Pacini. Rotating neutron stars, pulsars and supernova remnants. *Nature*, 219(5150):145–146, 1968. 3.1

[8] Thomas Gold. Rotating neutron stars as the origin of the pulsating radio sources. In *Pulsating Stars*, pages 74–75. Springer, 1968. 3.1

[9] Peter Goldreich and William H Julian. Pulsar electrodynamics. *The Astrophysical Journal*, 157:869, 1969. 3.1

[10] PA Sturrock. A model of pulsars. *The Astrophysical Journal*, 164:529, 1971. 3.1

[11] RN Manchester, GB Hobbs, A Teoh, and M Hobbs. The australia telescope national facility pulsar catalogue. *The Astronomical Journal*, 129(4):1993, 2005. 3.1

[12] Joris PW Verbiest and GM Shaifullah. Measurement uncertainty in pulsar timing array experiments. *Classical and Quantum Gravity*, 35(13):133001, July 2018. 3.1, 3.2

[13] AG Lyne, SL Shemar, and F Graham Smith. Statistical studies of pulsar glitches. *Monthly Notices of the Royal Astronomical Society*, 315(3):534–542, 2000. 3.1

[14] DC Backer, Shrinivas R Kulkarni, Carl Heiles, et al. A millisecond pulsar. *Nature*, 300(5893):615–618, 1982. 3.1

[15] Dipankar Bhattacharya and Edeard Peter Jacobus van den Heuvel. Formation and evolution of binary and millisecond radio pulsars. *Physics Reports*, 203(1-2):1–124, 1991. 3.1

[16] DR Lorimer and M Kramer. *Handbook of Pulsar Astronomy*. Cambridge University Press, Cambridge, UK, October 2012. 1

[17] DJ Helfand, RN Manchester, and JH Taylor. Observations of pulsar radio emission. iii-stability of integrated profiles. *The Astrophysical Journal*, 198:661–670, 1975. 3.2

[18] JPW Verbiest, JM Weisberg, AA Chael, et al. On pulsar distance measurements and their uncertainties. *The Astrophysical Journal*, 755(1):39, 2012. 3.2

[19] Timothy H Hankins and Barney J Rickett. Pulsar signal processing. In *Methods in Computational Physics: advances in research and applications*, volume 14, pages 55–129. Elsevier, 1975. 3.2

[20] Russell T Edwards, GB Hobbs, and RN Manchester. Tempo2, a new pulsar timing package–ii. the timing model and precision estimates. *Monthly Notices of the Royal Astronomical Society*, 372(4):1549–1574, 2006. 3.2, 3.2

[21] GB Hobbs, RT Edwards, and RN Manchester. Tempo2, a new pulsar-timing package–i. an overview. *Monthly Notices of the Royal Astronomical Society*, 369(2):655–672, 2006. 3.2

[22] G Hobbs, F Jenet, KJ Lee, et al. tempo2: a new pulsar timing package–iii. gravitational wave simulation. *Monthly Notices of the Royal Astronomical Society*, 394(4):1945–1955, 2009. 3.2

[23] Jing Luo, Scott Ransom, Paul Demorest, et al. PINT: A modern software package for pulsar timing. *The Astrophysical Journal*, 911(1):45, 2020. 3.2

[24] PW Anderson and N Itoh. Pulsar glitches and restlessness as a hard superfluidity phenomenon. *Nature*, 256(5512):25–27, 1975. 3.2

[25] Cristobal M Espinoza, Andrew G Lyne, Ben W Stappers, and Michael Kramer. A study of 315 glitches in the rotation of 102 pulsars. *Monthly Notices of the Royal Astronomical Society*, 414(2):1679–1704, 2011. 3.2

[26] G Hobbs, AG Lyne, and M Kramer. An analysis of the timing irregularities for 366 pulsars. *Monthly Notices of the Royal Astronomical Society*, 402(2):1027–1048, 2010. 3.2

[27] Andrew Lyne, George Hobbs, Michael Kramer, et al. Switched magnetospheric regulation of pulsar spin-down. *Science*, 329(5990):408–412, 2010. 3.2

[28] Ryan M Shannon and James M Cordes. Assessing the role of spin noise in the precision timing of millisecond pulsars. *The Astrophysical Journal*, 725(2):1607–1619, December 2010. 3.2

[29] KJ Lee, CG Bassa, GH Janssen, et al. Model-based asymptotically optimal dispersion measure correction for pulsar timing. *Monthly Notices of the Royal Astronomical Society*, 441(4):2831–2844, July 2014. 3.2

[30] MJ Keith, W Coles, RM Shannon, et al. Measurement and correction of variations in interstellar dispersion in high-precision pulsar timing. *Monthly Notices of the Royal Astronomical Society*, 429(3):2161–2174, March 2013. 3.2

[31] MV Sazhin. Opportunities for detecting ultralong gravitational waves. *Soviet Astronomy*, 22:36–38, 1978. 3.3

[32] Steven Detweiler. Pulsar timing measurements and the search for gravitational waves. *The Astrophysical Journal*, 234:1100–1104, 1979. 3.3

[33] Frank B Estabrook and Hugo D Wahlquist. Response of doppler spacecraft tracking to gravitational radiation. *General Relativity and Gravitation*, 6(5):439–447, 1975. 3.3

[34] William L Burke. Large-scale random gravitational waves. *The Astrophysical Journal*, 196:329–334, 1975. 3.3

[35] Michele Maggiore. *Gravitational Waves: Astrophysics and Cosmology*, volume 2. Oxford University Press, 2018. 3.3, 3.4

[36] Melissa Anholm, Stefan Ballmer, Jolien DE Creighton, et al. Optimal strategies for gravitational wave stochastic background searches in pulsar timing data. *Physical Review D*, 79(8):084030, April 2009. 3.4

[37] CMF Mingarelli, T Sidery, I Mandel, and A Vecchio. Characterizing gravitational wave stochastic background anisotropy with pulsar timing arrays. *Physical Review D*, 88(6):062005, September 2013. 3.4, 3.4

[38] Jonathan Gair, Joseph D Romano, Stephen Taylor, and Chiara MF Mingarelli. Mapping gravitational-wave backgrounds using methods from CMB analysis: Application to pulsar timing arrays. *Physical Review D*, 90(8):082001, October 2014. 3.4, 3.4, 3.4, 3.4

[39] Fredrick A Jenet and Joseph D Romano. Understanding the gravitational-wave Hellings and Downs curve for pulsar timing arrays in terms of sound and electromagnetic waves. *American Journal of Physics*, 83(7):635–645, July 2015. 3.4

[40] Chiara MF Mingarelli and Trevor Sidery. Effect of small interpulsar distances in stochastic gravitational wave background searches with pulsar timing arrays. *Physical Review D*, 90(6):062011, September 2014. 3.4

[41] RW Hellings and GS Downs. Upper limits on the isotropic gravitational radiation background from pulsar timing analysis. *The Astrophysical Journal*, 265:L39–L42, February 1983. 3.4, 3.3

[42] Elinore Roebber. Ephemeris errors and the gravitational-wave signal: Harmonic mode coupling in pulsar timing array searches. *The Astrophysical Journal*, 876(1):55, May 2019. 3.4, 3.4

[43] Elinore Roebber and Gilbert Holder. Harmonic space analysis of pulsar timing array redshift maps. *The Astrophysical Journal*, 835(1):21, January 2017. 3.4, 3.4

[44] Stephen R Taylor and Jonathan R Gair. Searching for anisotropic gravitational-wave backgrounds using pulsar timing arrays. *Physical Review D*, 88(8):084001, October 2013. 3.4

[45] SR Taylor, CMF Mingarelli, JR Gair, et al. Limits on anisotropy in the nanohertz stochastic gravitational wave background. *Physical Review Letters*, 115(4):041101, July 2015. 3.4

[46] Ciarán Conneely, Andrew H Jaffe, and Chiara MF Mingarelli. On the amplitude and Stokes parameters of a stochastic gravitational-wave background. *Monthly Notices of the Royal Astronomical Society*, 487(1):562–579, July 2019. 3.4

[47] Selim C Hotinli, James B Mertens, Matthew C Johnson, and Marc Kamionkowski. Probing correlated compensated isocurvature perturbations using scale-dependent galaxy bias. *Physical Review D*, 100(10):103528, November 2019. 3.4

[48] Stephen R Taylor, Rutger van Haasteren, and Alberto Sesana. From bright binaries to bumpy backgrounds: Mapping realistic gravitational wave skies with pulsar-timing arrays. *Physical Review D*, 102(8):084039, October 2020. 3.4

[49] Neil J Cornish and A Sesana. Pulsar timing array analysis for black hole backgrounds. *Classical and Quantum Gravity*, 30(22):224005, November 2013. 3.4

[50] Jonathan R Gair, Joseph D Romano, and Stephen R Taylor. Mapping gravitational-wave backgrounds of arbitrary polarisation using pulsar timing arrays. *Physical Review D*, 92(10):102003, November 2015. 3.4

[51] Maximiliano Isi and Leo C Stein. Measuring stochastic gravitational-wave energy beyond general relativity. *Physical Review D*, 98(10):104025, November 2018. 3.4

[52] Neil J Cornish, Logan O'Beirne, Stephen R Taylor, and Nicolás Yunes. Constraining alternative theories of gravity using pulsar timing arrays. *Physical Review Letters*, 120(18):181101, May 2018. 3.4

[53] Sydney J Chamberlin and Xavier Siemens. Stochastic backgrounds in alternative theories of gravity: Overlap reduction functions for pulsar timing arrays. *Physical Review D*, 85(8):082001, April 2012. 3.4

[54] KJ Lee, FA Jenet, and RH Price. pulsar timing as a probe of non-Einsteinian polarizations of gravitational waves. *The Astrophysical Journal*, 685:1304–1319, October 2008. 3.4

[55] Wenzer Qin, Kimberly K Boddy, and Marc Kamionkowski. Subluminal stochastic gravitational waves in pulsar-timing arrays and astrometry. *Physical Review D*, 103:024045, 2021. 3.4

[56] K Lee, FA Jenet, RH Price, et al. Detecting massive gravitons using pulsar timing arrays. *The Astrophysical Journal*, 722:1589–1597, October 2010. 3.4

CHAPTER 4

Sources & Signals

As in other portions of the GW spectrum, the overwhelming majority of signals in the PTA band are likely to be of compact-binary origin. Hence, I'll devote the bulk of this chapter to the dominant expected class of sources for PTAs: binary systems of supermassive black holes with masses $\sim 10^8$–$10^{10} M_\odot$. While these systems should be plentiful, teasing them apart from one another will be challenging to the PTA detector response and frequency resolution, rendering our first target to be the statistical signal aggregation over the entire population in the form of a stochastic GW background. The properties of this background encode demographic information of the binary systems, as well as details of their dynamical evolution in the final parsec of their journey to coalescence. Nevertheless template-based searches for individual binaries are possible, and appropriate waveform models are given here. Beyond binaries, I will give a brief overview of other potential PTA sources, including primordial GWs, GWs from cosmic strings and phase transitions, and finally other non-GW sources of inter-pulsar correlated timing delays in the form of timing systematics and even scalar-field dark matter.

4.1 SUPERMASSIVE BINARY BLACK HOLES

There is now general agreement that massive black holes (MBHs) reside at the centers of most galaxies (1; 2), with several well-known scaling relations indicating a symbiosis with the properties of the host galaxy (e.g., M–σ, M–M_{bulge} (3; 4; 5; 6; 7)). But conventional astronomical techniques are limited to studying them either locally or in quasar environments. The formation of MBH binaries should be a natural by-product of the hierarchical growth of galaxies through major and minor mergers (a process that also includes gas and dark matter accretion from cosmic web filaments) in ΛCDM cosmologies (8).

DOI: 10.1201/9781003240648-4

Indeed such MBH binaries (MBHBs) should be among the loudest GW sources in the Universe, and in the mass range $\sim 10^8 - 10^{10} M_\odot$ (referred to as *super*-massive black hole binaries; SMBHBs) constitute a key target population for PTAs. This requires the SMBHs to reach very close separations (on the order of milliparsecs) such that their GW emission would fall within the sensitive frequency range of PTAs ($\sim 1 - 100$ nHz). The chain of interactions that govern the inward migration of these black holes is discussed later in this chapter. The existing observational electromagnetic evidence for such sub-parsec separation SMBHBs is tenuous. There are many known "dual" supermassive black hole systems separated by kiloparsecs to hundreds of parsecs that are observed across X-ray, optical, and radio (see, e.g., 9, and references therein for a comprehensive review). The closest known directly resolved binary is at a projected separation of ~ 7 parsecs (10). Below a parsec, the evidence becomes indirect since telescopes lack the requisite spatial resolution to separate the binary components. Binary candidates are inferred through either apparent periodic photometric variability of AGN lightcurves (e.g., 11; 12; 13; 14), or large long-timescale velocity offsets in spectroscopically-derived AGN radial velocity curves (e.g., 15; 16; 17; 18; 19). Both are interpreted as effects of binarity through the interplay of gas and accretion disks surrounding each black hole and the binary as a whole.

4.1.1 Characteristic Strain Spectrum

We saw in Chapter 2 that the energy density in GWs is defined as

$$\Omega_{\text{SGWB}}(f) \equiv \frac{1}{\rho_c} \frac{d\rho}{d \ln f}. \tag{4.1}$$

where f is the GW frequency, ρ is the GW energy density, and ρ_c is the closure density that corresponds to flat cosmological geometry. I will first demonstrate a simple back-of-the-envelope scaling relationship for the GWB energy density in a population of circular inspiraling compact binary systems, which can then be trivially converted into a characteristic strain spectrum. While this derivation is specifically in the context of supermassive binary black holes (the major subject of this chapter), the derivation applies to any compact binary population. We write the term $d\rho/d \ln f$ as an integral over a continuous distribution of emitting sources:

$$\frac{d\rho}{d \ln f} = \int_0^\infty dz \frac{dn}{dz} \frac{1}{(1+z)} \frac{dE}{d \ln f_r} \bigg|_{f_r = f(1+z)} \tag{4.2}$$

where dn/dz is the number density distribution of binaries over redshift, the factor of $1/(1+z)$ accounts for the cosmological redshifting of GW energy, and $dE/d \ln f_r$ is the *source-frame* GW energy emitted by the binary per logarithmic frequency interval. Note that f_r refers to the source-frame (or "rest-frame") frequency. The interpretation of this is simple: within a given logarithmic frequency interval, we are adding together the energy from all binaries throughout the Universe whose emission results in source-frame frequencies redshifted into the relevant observed frequency range. By using scaling relations given in Chapter 2, 2.2.2.1, we can write the GW energy spectrum as

$$\frac{dE}{d\ln f_r} = f_r \frac{dE}{dt_r}\frac{dt_r}{df_r} \propto f_r \times f_r^{10/3} \times f_r^{-11/3} \propto f_r^{2/3}. \tag{4.3}$$

The source-frame GW frequency is simply the redshift-corrected observed frequency, $f_r = f(1 + z)$, and we have seen that the frequency of GW emission from a circular binary system is simply twice the orbital frequency. Plugging these scaling relationships into $d\rho/d\ln f$ and collecting frequency terms gives us $\Omega_{\text{SGWB}}(f) \equiv (1/\rho_c)(d\rho/d\ln f) \propto f^{2/3}$. Finally, using Eq. 2.27 we can write that the characteristic strain spectrum from a population of circular inspiraling compact binaries is (20; 21; 22; 23)

$$h_c(f) \propto f^{-2/3}. \tag{4.4}$$

As mentioned, in PTA searches for a SGWB the relevant compact binary population consists of supermassive black holes. Given that we are probing a GW frequency range of $\sim$1–100 nHz, historical convention opts for a reference frequency of $f_{\text{ref}} = 1\text{yr}^{-1}$,[1] such that

$$h_c(f) = A_{\text{SGWB}} \left(\frac{f}{1\text{yr}^{-1}}\right)^\alpha \tag{4.5}$$

where $\alpha = -2/3$ for our population of circular inspiraling supermassive binary black holes.

Let's now look more carefully at how we build up the characteristic strain spectrum from a population of SMBHBs. We will relax our previous assumption of circularity, and also retain all dependencies on the distribution of system properties. This discussion closely follows Ref. (24) and references therein; we refer the reader there for further details. The squared characteristic strain spectrum can be written as

$$h_c^2(f) = \int_0^\infty \int_0^\infty \int_0^1 dz dM_1 dq \frac{d^4N}{dz dM_1 dq dt_r} \times$$
$$\sum_{n=1}^\infty \left\{ \frac{g[n, e(f_{K,r})]}{(n/2)^2} \frac{dt_r}{d\ln f_{K,r}} h^2(f_{K,r}) \right\}, \tag{4.6}$$

where M_1 is the mass of the primary BH; $0 < q \le 1$ is the binary mass ratio (M_2/M_1); and $f_{K,r}$ is the source-frame Keplerian orbital frequency of the binary, where the GW emission from eccentric binaries will result in redshifted harmonics of this, $f_n = nf_{K,r}/(1 + z)$. The other terms of the integrand split into population factors and individual system factors. The term

---

[1]PTAs actually have terrible sensitivity at this frequency because of the need to fit for the pulsar's sky location in the timing model, but this is ultimately irrelevant since most of the signal significance derives from the lowest frequencies in the band regardless of what the model's reference frequency is.

$d^4N/dz dM_1 dq dt_r$ is the population weighting of each binary's contribution to the squared strain, corresponding to the comoving merger rate of SMB-HBs per redshift, primary mass, and mass-ratio interval, where t_r measures source-frame time. The remaining terms are all related to the contribution of each SMBHB to the squared strain. As noted above, the GW emission from an eccentric binary will be distributed over harmonics of the orbital frequency (25). We are thus summing up contributions of the GW emission from each binary into redshifted harmonics of their orbital frequency that co-incide with the relevant observed GW frequency. The function $g(n, e)$ is an eccentricity-dependent function describing the distribution of GW emission into orbital-frequency harmonics (25). In the case of a circular binary with $e = 0$, we have $g(n, e = 0) = 0$ for $n \neq 2$, resulting in our much simpler derivation above. The term $h(f_{K,r})$ is defined as

$$h(f_{K,r}) = \sqrt{\frac{32}{5}} \frac{\mathcal{M}^{5/3} (2\pi f_{K,r})^{2/3}}{D_c}, \qquad (4.7)$$

which corresponds to the orientation-averaged GW strain amplitude of a single SMBHB, where $\mathcal{M} := (M_1 M_2)^{3/5}/(M_1 + M_2)^{1/5}$ is known as the "chirp mass" (in the source frame), and D_c is the radial comoving distance to the binary system.

Finally, the term $dt_r/d\ln f_{K,r}$ describes the amount of time that each SMBHB spends emitting in a particular logarithmic $f_{K,r}$ interval. If the binary is evolving purely through GW emission then this follows known relationships that account for enhancement in the rate of evolution due to eccentricity (25; 26). However, at wider separations the binaries may remain in contact with their ambient astrophysical environment, and their orbital evolution may be partially driven by these factors. Hence, a more general description of this term sums over all processes that may be influencing the orbital evolution of a given SMBHB:

$$\frac{dt_r}{d\ln f_{K,r}} = f_{K,r} \times \left[\sum_k \frac{df_{K,r}}{dt_r}\right]^{-1}. \qquad (4.8)$$

4.1.2 Binary Dynamical Evolution

The coalescence of two SMBHs occurs after a chain of dynamical interactions that begins with the merger of their host galaxies. Several of these interactions can imprint themselves on the SGWB characteristic strain spectrum through their influence on $dt_r/\ln f_{K,r}$, yielding an opportunity for PTAs to constrain the ensemble sub-parsec dynamical behavior of SMBHBs. The net result is that the strain spectrum can be attenuated at low frequencies and de-part from the fiducial $f^{-2/3}$ behavior to give a *turnover*, since these effects accelerate the evolution of the SMBHB orbit at wider separations. Ref. (27) suggested a simple parametrization of the strain spectrum for a population of circular environmentally-influenced binaries:

$$h_c(f) = A_{\text{SGWB}} \frac{\left(f/1\text{yr}^{-1}\right)^{-2/3}}{[1 + (f_{\text{bend}}/f)^{\kappa}]^{1/2}}, \tag{4.9}$$

where f_{bend} is the orbital frequency below which environmental interactions dominate the SMBHB evolution, and κ is determined by the dominant dynamical interaction. We will see the values that κ can take as each interaction is introduced below. As discussed later in Chapter 7, there are a variety of other astrophysically-driven spectral models of the SGWB in use (28; 29; 30; 31). A range of possible SGWB shapes due to varying astrophysical conditions and dynamical interactions are shown in Fig. 4.1, along with some PTA strain upper limits for orientation.

4.1.2.1 Dynamical Friction

Dynamical friction is the term describing a viscous drag influence on SMBHs in the environment of a galactic merger. After this merger, the accumulated effect of many weak and long-range gravitational scattering events brings the "dual" system – composed of the respective SMBHs and stellar cores – from ~kpc separations down to ~1 parsec (34; 35; 36; 37; 33; 38). For a SMBH spiraling toward the center of a spherical galaxy on a circular orbit, the inspiral timescale is of the order of Gyrs (39; 38). Nevertheless, dynamical friction is an important initial step in reducing the separation of the two black holes from kiloparsecs. It does not yield a measurable influence on the shape of the SGWB strain spectrum.

4.1.2.2 Stellar Loss-cone Scattering

At parsec separations, the individual black holes are moving fast enough that dynamical friction no longer exerts an influential tightening drag. The dominant mechanism of binary hardening then results from individual scattering events between low angular-momentum stars in the galactic core and the SMBHB (40; 41; 42; 43). The *loss cone* (LC) (41) describes the region of stellar-orbit phase space where orbits are centrophilic enough to scatter off the binary. The original definition of the *final parsec problem* (44; 45) referred to the potential depletion of this phase-space reservoir, resulting in stalled SMBHBs. In practice the triaxiality and rotation of post-merger galaxies can ensure a healthy supply of such stars (46; 47; 48; 49). The scattering and ejection of stars by the binary leads to hardening through the following semi-major axis and eccentricity evolution (43):

$$\frac{da}{dt} = -\frac{G\rho}{\sigma} H a^2, \quad \frac{de}{dt} = \frac{G\rho}{\sigma} H K a, \tag{4.10}$$

where $H \sim 15$ is a dimensionless hardening rate, and $K \sim 0.1$ is a dimensionless eccentricity *growth* rate, i.e., stellar LC scattering can promote eccentricity growth. Both of these can be computed from numerical scattering experiments

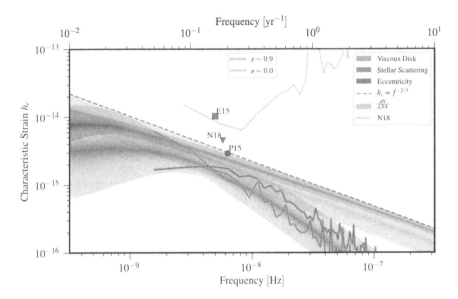

Figure 4.1: An example of the characteristic strain spectrum of a SGWB of SMBHB-origin, as a function of GW frequency in Hz (bottom axis) and inverse years (top axis). The dashed line shows the fiducial $f^{-2/3}$ spectrum. The influences of dynamical interactions are shown as bands: stellar losscone scattering leading onto viscous disk coupling is shown as a thinning band becoming closer to $f^{-2/3}$ as frequency increases. The wide band with a pronounced turnover at low frequencies shows an ensemble of spectra from eccentric SMBHB populations. The solid lines show individual realizations of high and low population-eccentricity spectra, where the pronounced turnover spectrum has binaries with initial eccentricity of 0.9. Figure reproduced with permission from Ref. (24) using populations based on Refs. (32; 33).

for specific mass ratio and eccentricity conditions (50). If we take stellar LC scattering to be the dominant term in $dt_r/\ln f_{K,r}$, then the resulting strain spectrum behavior is $h_c(f) \propto f$, which is clearly very different from the fiducial $\propto f^{-2/3}$ circular GW-driven behavior. This corresponds to $\kappa = 10/3$ in Equation 4.9. Eccentricity growth will further attenuate the strain spectrum at low frequencies (e.g., 51; 32).

4.1.2.3 Viscous Circumbinary Disk Interaction

At centiparsec to milliparsec separations, *viscous dissipation of orbital energy to a gaseous circumbinary disk* could play a vital role in hardening the binary (40; 52; 53; 54). The binary torque will dominate over the viscous torque in the disk, leading to the formation of a cavity in the gas distribution and the accumulation of material at the outer edge of this cavity (i.e., Type II migration). The excitation of a spiral density wave in the disk torques the binary, and leads to hardening through the following semi-major axis evolution (53; 54):

$$\frac{da}{dt} = -\frac{2\dot{M_1}}{\mu}(aa_0)^{1/2}, \qquad (4.11)$$

where $\dot{M_1}$ is the mass accretion rate onto the primary BH, μ is the binary reduced mass, and a_0 is the semi-major axis at which the disk mass enclosed is equal to the mass of the secondary BH (53).

As studied comprehensively in Ref. (52), the strain spectrum under different disk–binary scenarios can vary from $h_c(f) \propto f^{-1/6}$ ($\kappa = 1$ in Equation 4.9), to $h_c(f) \propto f^{1/2}$ for the model in Equation 4.11 ($\kappa = 7/3$ in Equation 4.9). Across all models, the characteristic strain spectrum can be flattened or even have positive slope due to disk coupling. There are many caveats to these simple spectral parametrizations, and the actual spectral shape will depend on the detailed dissipative physics of the disk, and the disk-binary dynamics. In fact, as of writing there is emerging evidence of viscous disk interaction actually *widening* the binary separation (i.e., da/dt has a positive sign) under certain conditions (e.g., 55; 56; 57).

4.1.2.4 Gravitational-wave Inspiral

Once the binary decouples from its astrophysical environment at the smallest scales ($\lesssim$ milliparsec), the emission of gravitational radiation will dominate the orbital evolution. The dissipation of orbital energy then only depends on the binary component masses, the orbital semi-major axis, and the eccentricity. We have already seen that this leads to $h_c \propto f^{-2/3}$. However, this is an ensemble averaged behavior, and finiteness in a given realization of the emitting SMBHB population will lead to departures from this power-law spectrum at frequencies beyond $\sim 10^{-8}$ Hz, tilting it more steeply (58).

But, what happens if LC scattering and disk interactions don't bring the SMBHs close enough to actually coalesce within a Hubble time? What if all binaries stall before they can create a measurable signal in the relevant range of PTA-sensitive GW frequencies? A third mechanism may come to the rescue: *triplet interactions.* Galaxies can undergo numerous merger events over cosmic time (e.g., 59), with the potential to form hierarchical MBH systems that undergo Kozai-Lidov oscillations (60; 61), thereby driving up the eccentricity of the inner binary and accelerating coalescence due to GW emission (62; 63; 64; 65; 66; 67; 68). Eccentricity itself can leave an imprint on the shape of the strain spectrum (69; 70; 71; 32; 72; 51), where its effect is to evolve each system more rapidly (thereby reducing the occupation fraction of frequency bins), and to distribute GW strain across higher orbital frequency harmonics. The result is that the spectrum can exhibit a flattening/turnover at low frequencies, a small enhancement at the turnover transition, and then a return to the usual $f^{-2/3}$ behavior at the highest frequencies (where the population will be mostly circularized).

4.1.3 Signal from an Individual Binary

Equation 3.13 showed the GW-induced fractional shift to the pulse arrival rate, which took the form

$$z(t, \hat{\Omega}) = \frac{1}{2} \frac{p^a p^b}{(1 + \hat{\Omega} \cdot \hat{p})} \Delta h_{ab}, \tag{4.12}$$

where Δh_{ab} is the difference in the spatial components of the metric perturbation between the time at which the GW passes the Earth and when it passed the pulsar with positional unit vector $\hat{p}$. Previously we had only been interested in the statistical properties of this quantity for the purposes of describing a SGWB. However, for a single SMBBH we can write a determinstic waveform model to be deployed in searches for individual sources of GWs. In the TT gauge we write the metric perturbation for a GW propagating in direction $\hat{\Omega}$ at an arbitrary time as

$$h_{ab}(t, \hat{\Omega}) = h_+(t)e^+_{ab}(\hat{\Omega}) + h_\times(t)e^\times_{ab}(\hat{\Omega}). \tag{4.13}$$

The polarization basis tensors can be defined irrespective of the binary properties in terms of a basis triad:

$$e^+_{ab} = \hat{u}_a \hat{u}_b - \hat{v}_a \hat{v}_b,$$
$$e^\times_{ab} = \hat{u}_a \hat{v}_b + \hat{v}_a \hat{u}_b, \tag{4.14}$$

$$\hat{n} \equiv -\hat{\Omega} = (\sin\theta\cos\phi, \sin\theta\sin\phi, \cos\theta),$$
$$\hat{u} = (\cos\psi\cos\theta\cos\phi - \sin\psi\sin\phi,$$
$$\cos\psi\cos\theta\sin\phi + \sin\psi\cos\phi, -\cos\psi\sin\theta),$$
$$\hat{v} = (\sin\psi\cos\theta\cos\phi + \cos\psi\sin\phi,$$
$$\sin\psi\cos\theta\sin\phi - \cos\psi\cos\phi, -\sin\psi\sin\theta), \tag{4.15}$$

where $(\theta, \phi) = (\pi/2 - \text{DEC}, \text{RA})$ denotes the sky-location of the binary in spherical polar coordinates, and ψ is the GW polarization angle that corresponds to the angle between $\hat{u}$ and the line of constant azimuth when the orbit is viewed from our coordinate system origin. See the right panel of Fig. 4.2.

For the polarization amplitudes, I'll adopt the less conventional approach of tackling arbitrary binary eccentricity first, then showing how the model reduces down to its circular form. This discussion closely follows Ref. (73), which employs leading-order Peters & Matthews waveforms (25) that are described in Ref. (74). I'll be ignoring the influence of higher order post-Newtonian terms, although readily-usable waveforms inclusive of these do exist for PTA analysis, and are being actively implemented (75). This discussion also ignores BH spin, whose observable influences are likely to be challenging for PTAs to infer within the next decade (76; 77). The relevant polarization amplitudes for the metric perturbation can be expressed analytically as

$$h_+(t) = \sum_n -(1 + \cos^2 \iota)[a_n \cos(2\gamma) - b_n \sin(2\gamma)] + (1 - \cos^2 \iota)c_n,$$
$$h_\times(t) = \sum_n 2 \cos \iota [b_n \cos(2\gamma) + a_n \sin(2\gamma)], \tag{4.16}$$

where formally the summation is over integers $n = [1, \ldots, \infty]$, but in practice can be truncated (73), and

$$a_n = -n\zeta\omega^{2/3} [J_{n-2}(ne) - 2eJ_{n-1}(ne) + (2/n)J_n(ne)$$
$$+ 2eJ_{n+1}(ne) - J_{n+2}(ne)] \cos[nl(t)],$$
$$b_n = -n\zeta\omega^{2/3}\sqrt{1 - e^2} [J_{n-2}(ne) - 2J_n(ne) + J_{n+2}(ne)] \sin[nl(t)],$$
$$c_n = 2\zeta\omega^{2/3} J_n(ne) \cos[nl(t)]. \tag{4.17}$$

The parameters in these coefficients are defined as follows:

- The amplitude is $\zeta = \mathcal{M}_z^{5/3}/D_L$, where $D_L = (1+z)D_c$ is the luminosity distance of the binary, and $\mathcal{M}_z = (1+z)\mathcal{M}$ is the *redshifted* chirp mass.

- The quantity $l(t)$ is known as the *mean anomaly*, defined by $l(t) - l_0 = 2\pi \int_{t_0}^t f_K(t')dt'$, where $f_K = f_{K,r}/(1+z)$ is the observer-frame Keplerian orbital frequency, and $\omega = 2\pi f_K$. The mean anomaly is essentially the orbital phase of the binary if we had been dealing with a circular system (see the left panel of Fig. 4.2).

- The angle γ is an azimuthal angle measuring the direction of the system pericenter with respect to $\hat{x} \equiv (\hat{\Omega} + \hat{\mathcal{L}} \cos \iota)/\sqrt{1 - \cos^2 \iota}$, where $\hat{\mathcal{L}}$ is a unit vector pointing along the binary's orbital angular momentum.

- The binary orbital inclination angle, ι, is defined by $\cos \iota = -\hat{\mathcal{L}} \cdot \hat{\Omega}$, i.e., $\iota = \pi/2$ corresponds to an edge-on binary.

- The parameter $0 < e < 1$ is the eccentricity, describing the ellipticity of the system, where a perfectly circular orbit would have $e = 0$.

- The functions $J_{(\cdot)}(\cdot)$ are Bessel functions.

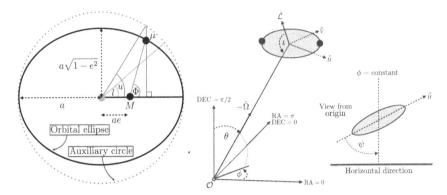

Figure 4.2: *Left:* Orbital geometry of an eccentric binary. The motion is shown as the binary reduced mass, $M_1 M_2/(M_1+M_2)$, orbiting the total mass, $(M_1 + M_2)$, which is positioned at a focus of the ellipse. The semi-major axis is a, the eccentricity is e, the orbital phase is Φ, the mean anomaly is l, and the eccentric anomaly is u. *Right:* The orientation of the binary orbit with respect to the observer's coordinate system, where $\hat{n} = -\hat{\Omega}$, $\hat{u}$, and $\hat{v}$ form a basis triad with which to describe the GW polarization basis tensors, $\hat{\mathcal{L}}$ is a unit vector in the direction of the orbital angular momentum, ψ is the GW polarization angle, and $\{\theta, \phi\}$ are the usual spherical-polar coordinates. Both figures reproduced from Ref. (73).

This Fourier solution to Kepler's problem for a binary system makes explicit that the GW radiation from an eccentric binary is not monochromatic, and in fact occurs over a spectrum of harmonics of the orbital frequency. However, as expected, when $e = 0$ we have Bessel terms $J_0(0) = 1$ and $J_{n>0}(0) = 0$, such that the only remaining GW emission occurs at $f = 2f_{K,r}/(1 + z)$.

Finally, we can write our model for the pulsar timing delays that are induced by an individual SMBBH. As previously introduced, this is the integrated effect over the induced fractional shift to the pulse arrival rate:

$$R(t) \equiv \int_0^t dt' \, z(t') = F^+(\hat{\Omega})\Delta s_+(t) + F^\times(\hat{\Omega})\Delta s_\times(t), \qquad (4.18)$$

where $\Delta s_{+,\times}$ are the corresponding differences in Earth-term and pulsar-term delays, and $F^{+,\times}$ are the previously introduced GW antenna response patterns for each polarization (see Eq. 2.34). The time-dependent components of $s_{+,\times}$ can be computed analytically under the assumption that the system is non-evolving over the observational baseline of the pulsar (on average $\sim 10 - 20$ years) at the time when the GW passes the Earth (for the Earth-term) and at the time at which the GW previously passed the relevant pulsar (for the pulsar-term). This means that quantities of time only appear linearly in the definition of the mean anomaly. In a modification to the description in Ref. (73), we allow for evolution of the pericenter angle γ over the observation time (74), for which we denote

$$\dot{\gamma} \equiv \frac{d\gamma}{dt} = 3\omega \frac{(\omega M)^{2/3}}{(1-e^2)} \left[1 + \frac{(\omega M)^{2/3}}{4(1-e^2)}(26 - 15e^2)\right]. \tag{4.19}$$

Thus $\gamma(t) \approx \gamma_0 + \dot{\gamma}t$, and $l(t) \approx l_0 + \omega t$. Notice that the pericenter will advance even for a perfectly circular orbit. Therefore at a given time, either during the baseline of the Earth- or pulsar-term, the time dependence of the GW-induced delay from a single SMBBH is given by

$$s_+(t) = \sum_n -(1 + \cos^2 \iota)[\tilde{a}_n A_n(t) - \tilde{b}_n B_n(t)] + (1 - \cos^2 \iota)\tilde{c}_n,$$
$$s_\times(t) = \sum_n 2 \cos \iota [\tilde{a}_n C_n(t) - \tilde{b}_n D_n(t)], \tag{4.20}$$

where

$$A_n(t) = \frac{1}{2}\left[\frac{\sin[nl(t) - 2\gamma(t)]}{n\omega - 2\dot{\gamma}} + \frac{\sin[nl(t) + 2\gamma(t)]}{n\omega + 2\dot{\gamma}}\right],$$
$$B_n(t) = \frac{1}{2}\left[\frac{\sin[nl(t) - 2\gamma(t)]}{n\omega - 2\dot{\gamma}} - \frac{\sin[nl(t) + 2\gamma(t)]}{n\omega + 2\dot{\gamma}}\right],$$
$$C_n(t) = \frac{1}{2}\left[\frac{\cos[nl(t) - 2\gamma(t)]}{n\omega - 2\dot{\gamma}} - \frac{\cos[nl(t) + 2\gamma(t)]}{n\omega + 2\dot{\gamma}}\right],$$
$$D_n(t) = \frac{1}{2}\left[\frac{\cos[nl(t) - 2\gamma(t)]}{n\omega - 2\dot{\gamma}} + \frac{\cos[nl(t) + 2\gamma(t)]}{n\omega + 2\dot{\gamma}}\right], \tag{4.21}$$

and

$$\tilde{a}_n = -n\zeta\omega^{2/3}[J_{n-2}(ne) - 2eJ_{n-1}(ne) + (2/n)J_n(ne)$$
$$+ 2eJ_{n+1}(ne) - J_{n+2}(ne)]$$
$$\tilde{b}_n = -n\zeta\omega^{2/3}\sqrt{1-e^2}[J_{n-2}(ne) - 2J_n(ne) + J_{n+2}(ne)]$$
$$\tilde{c}_n = (2/n)\zeta\omega^{-1/3}J_n(ne)\sin[nl(t)]. \tag{4.22}$$

This all seems quite complicated. But this is *almost* as easy as it gets, since we have ignored binary evolution during the Earth- and pulsar-term baselines, as well as higher post-Newtonian terms and spin contributions. From a modeler's perspective, we are fortunate that PTAs are likely to detect binaries that are in the early adiabatic inspiral regime of their path to coalescence. I said this is *almost* as easy as it gets, because of course all this simplifies significantly when $e = 0$ and we ignore pericenter advance. The term $\tilde{c}_n = 0$ $\forall n$, and only $\tilde{a}_2$ and $\tilde{b}_2$ are non-zero, taking values of $\tilde{a}_2 = \tilde{b}_2 = -2\zeta\omega^{2/3}$. Therefore, the

$s_{+,\times}$ terms reduce to

$$s_+(t) = \zeta\omega^{-1/3}(1 + \cos^2 \iota) \sin(2l_0 + 2\gamma_0 + 2\omega t)$$
$$= \zeta\omega^{-1/3}(1 + \cos^2 \iota) \sin(2\Phi_0 + 2\omega t),$$
$$s_\times(t) = 2\zeta\omega^{-1/3} \cos \iota \cos(2l_0 + 2\gamma_0 + 2\omega t)$$
$$= 2\zeta\omega^{-1/3} \cos \iota \cos(2\Phi_0 + 2\omega t), \tag{4.23}$$

where typically the angular terms are packaged together as $\Phi_0 \equiv l_0 + \gamma_0$, rendering an even more compact expression.

Nevertheless even in the circular, static-pericenter scenario, there is some orbital evolution that we can not ignore; this is the evolution that occurs between the stage of the pulsar-term (earlier) and the Earth-term (always later). This time difference can be thousands of years. For an array of pulsars, the Earth-terms are at a common stage of source dynamics, whereas the pulsar terms are snapshots of the orbital dynamics that lag behind the Earth by $t_e - t_p = L_p(1 + \hat{\Omega} \cdot \hat{p})$, i.e., they are pulsar specific. The orbital parameters of the source at the time of the pulsar-terms can be calculated by evolving the relevant Earth-term parameters backward in time using the evolution equations

$$\frac{d\omega}{dt} = \frac{96}{5\mathcal{M}^2}(\omega\mathcal{M})^{11/3}\frac{1 + \frac{73}{24}e^2 + \frac{37}{96}e^4}{(1 - e^2)^{7/2}},$$
$$\frac{de}{dt} = -\frac{304}{15\mathcal{M}}(\omega\mathcal{M})^{8/3}e\frac{1 + \frac{121}{304}e^2}{(1 - e^2)^{5/2}}, \tag{4.24}$$

where we see that GW emission causes a binary's orbital frequency to increase and its eccentricity to decrease. As the eccentricity decreases, the frequency of peak emitted GW power shifts to lower harmonics of the orbital frequency, eventually settling on $n = 2$ at $e = 0$. Eccentricity and orbital frequency co-evolve independently of the system mass (25; 78); a binary with an orbital frequency of 1 nHz and eccentricity of $e = 0.95$ will have partially circularized to $e \approx 0.3$ by the time that its orbital frequency has evolved to 100 nHz. Calculating the backwards evolution of the Earth-term parameters can either be done numerically, or (for circular systems), analytically:

$$\omega(t_p) = \omega_e\left(1 - \frac{256}{5}\mathcal{M}^{5/3}\omega_e^{8/3}t_p\right)^{-3/8}, \tag{4.25}$$

where subscripts e/p denote quantities at the time of Earth- or pulsar-terms. In fact, for weakly evolving systems we can simply perform a Taylor expansion around the Earth-term quantities such that

$$\omega(t_p) = \omega_e + \left.\frac{d\omega}{dt}\right|_e \times (t_p - t_e) = \omega_e - \left.\frac{d\omega}{dt}\right|_e \times L_p(1 + \hat{\Omega} \cdot \hat{p}). \tag{4.26}$$

It is essential to include the pulsar-terms for accurate inference of SMBHB

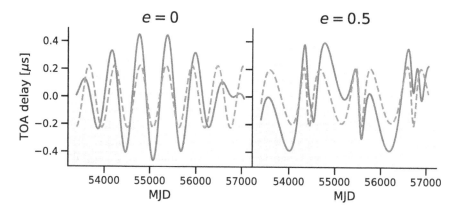

Figure 4.3: Example GW-induced TOA-delay time series for a circular binary ($e = 0$) and an eccentric binary ($e = 0.5$), over a baseline of 10 years. The blue dashed line shows the Earth term, while the red solid line shows the entire signal that includes the pulsar term. The pulsar is assumed to be 1 kpc distant, and in the sky location of PSR J1713+0747. The binary GW source has total mass of $3 \times 10^9 M_\odot$, $q = 1$, orbital frequency of 10 nHz, and at a distance of 20 Mpc.

properties from PTA data. Ignoring them can, at the very least, lead to biased recovery of the source's sky location, rendering the difficult problem of localization even more challenging (79; 80). Indeed, for sources that do not evolve during the observational baseline of the PTA, it is not possible to disentangle the source's chirp mass and distance without leveraging the pulsar-term information. These different snapshots separated by thousands of years over-constrain the problem, and in principle allow amplitude terms to be split apart for mass and distance recovery. In practice, modeling the pulsar-terms in Bayesian single-source searches is incredibly challenging; the likelihood can be highly oscillatory in the pulsar distances (which must be searched over) (79; 80; 81), and current prior bounds on these distances are not where we need them to be for useful inference; ideally these would be tighter than a gravitational wavelength, which corresponds to ~1 parsec for a source with GW frequency of 10 nHz. Some examples of TOA-delay time series caused by GWs from a circular and eccentric SMBHB are shown in Fig. 4.3.

4.1.4 Gravitational-wave Memory Burst

Before moving away from SMBHBs entirely, it bears mentioning that there is another important signature that they can impart. PTAs are not sensitive to the final oscillatory gravitational waveform signature of the binary coalescence; this is too high in frequency, lying somewhere between ~1 − 100 μHz (i.e., the gap between the high end of PTAs and the low end of LISA) for the relevant binary masses. However, they are sensitive to the associated broadband *burst with memory* (BWM) (82; 83; 84; 85). There are two flavors of GW memory: (i) *linear memory* that arises from non-oscillatory motion of the source system, particularly unbound masses (e.g., hyperbolic orbits, mass ejection), and (ii) *nonlinear memory* that is a direct consequence of the nonlinear structure of General Relativity, where GWs themselves source other GWs. We are specifically interested in nonlinear GW memory bursts in binary systems (86; 87), where during the final coalescence the memory effect leads to a permanent offset in the baseline of the GW oscillations. This residual deformation in spacetime has a rise-time of ∼1 day for a merger resulting in a $10^9 M_\odot$ BH (88; 89). This is shorter in duration than the typical pulsar-timing observation cadence, meaning that for all intents and purposes we can model the GW memory burst as a step-function in strain (modulated by antenna response considerations), and thus a ramp-function in the induced timing delays. Since this kind of ramp feature is broadband, BWM signals provide an indirect probe of the final merger of SMBHs (90; 88; 89).

The signal model for a BWM in PTA searches is

$$s_a(t) = h_{\text{mem}} B_a(\theta, \phi, \psi) \times [(t - t_0)\Theta(t - t_0) - (t - t_a)\Theta(t - t_a)], \quad (4.27)$$

where $B_a(\theta, \phi, \psi)$ describes the GW response of pulsar a in terms of its sky location (θ, ϕ) and GW polarization ψ; t_0 is the time at which the GW wavefront passes the Earth, t_a is the retarded time at which the wavefront passed the pulsar, $t_a = t_0 - L_a(1 + \hat{\Omega} \cdot \hat{p}_a)$, and Θ is the Heaviside step function. The memory strain is defined as

$$h_{\text{mem}} = \frac{1}{24} \frac{\eta M}{D_c} \sin^2 \iota (17 + \cos^2 \iota) \left[\frac{\Delta E_{\text{rad}}}{\eta M} \right],$$

$$\frac{\Delta E_{\text{rad}}}{\eta M} = 1 - \frac{\sqrt{8}}{3} \sim 0.06, \quad (4.28)$$

where M is the binary total mass, $\eta = M_1 M_2 / M^2$ is the reduced mass ratio, D_c is the radial comoving distance, ι is the inclination angle, and ΔE_{rad} is the energy radiated from the system. NANOGrav, the European Pulsar Timing Array, and the Parkes Pulsar Timing Array have all searched for BWM signals, and in the absence of detection have constrained the strain amplitude as a function of sky location and burst epoch (90; 91; 92; 93).

4.2   EXOTIC GRAVITATIONAL WAVE SOURCES

While SMBHBs form the likely initial source class for PTAs, there may be additional exotic sources of GWs lurking beneath the binary signals. These present a tantalizing probe of fundamental physics and the Universe at the earliest times.

4.2.1 Relic GWs

Electromagnetic observations of the Universe can not peer through the veil of the last-scattering surface, rendering conventional astronomy limited to times later than $380,000$ years after the Big Bang. However, the fact that GWs couple so weakly to matter (which makes their detection technologically challenging) also means that they can propagate throughout the Universe essentially unhindered.

If the Universe underwent an early period of exponential expansion (*inflation*), microscopic phenomena should have been rapidly amplified to macroscopic scales. This applies to quantum fluctuations in early spacetime, where inflation caused superadiabatic amplification of zero-point quantum fluctuations in the early gravitational field, leading to the production of a broadband background of primordial GWs (94; 95; 96). These GWs are a major target for polarization studies of the CMB, wherein excess "curl"-mode polarization could be the tell-tale signature (97, and references therein).

A simple model of the primordial/relic GW background can be expressed as a power-law in characteristic strain, where $h_c(f) = A(f/1\mathrm{yr}^{-1})^\alpha$, and $\alpha = n_t/2 - 2/(3w + 1)$ such that n_t is the tensor index (which depends on the detailed dynamics of inflation), and $w = p/\rho$ is the equation of state in the immediate post-inflation (but pre-BBN) Universe (98). For a scale-invariant primordial power spectrum ($n_t = 0$), and a radiation-dominated equation of state ($w = 1/3$), this predicts $\alpha = -1$ and $\gamma \equiv 3 - 2\alpha = 5$ (99; 100).

4.2.2 Cosmological Phase Transitions

As the Universe cools and expands, the material within it may undergo a first-order phase transition if its temperature drops below the characteristic temperature for the transition (101). Such phase transitions involve a discontinuity in first derivatives of the thermodynamic free energy, and are associated with the transitions we encounter in everyday life, e.g., solid/liquid/gas material transitions. Generally, they occur when the true minimum of a potential in the new phase is separated from a false minimum (i.e., the minimum from the previous phase) by a potential barrier through which a field must locally tunnel. In a cosmological scenario, bubbles of the new phase ("true vaccua") nucleate within the old phase ("false vaccuum"), expanding rapidly until the bubble walls move relativistically. Collisions between these bubble walls should be copious sources of GWs (102; 103; 104). Indeed, these bubble collisions precipitate phenomena that act as further sources of gravitational radiation, namely collisions of the sound waves that were generated within the false vaccuum by the original bubble wall collisions, and plasma turbulence generated by the expansion and collision of the sound waves.

The spectrum of GWs generated by first-order phase transitions does not obey a power-law shape, but does encode dependencies on the underlying generating processes. For example, the peak frequency of the spectrum is linearly related to the temperature T_*, and the bubble nucleation rate β_*, at the time of the phase transition (105; 106):

$$f_0^{\mathrm{peak}} \simeq 0.113\,\mathrm{nHz} \left(\frac{f_*^{\mathrm{peak}}}{\beta_*} \right) \left(\frac{\beta_*}{H_*} \right) \left(\frac{T_*}{\mathrm{MeV}} \right) \left(\frac{g_*}{10} \right)^{1/6}, \qquad (4.29)$$

where f_0^{peak} is the peak frequency today, f_*^{peak} is the peak frequency at the time of emission, H_* is the Hubble expansion rate at the time of emission, and g_* is the number of relativistic degrees of freedom. The full shape of the GW spectrum will depend on the interplay of the three sourcing phenomena. While the typical frequencies probed by PTAs are far below the Standard Model electroweak scale ($\lesssim 100$ Gev), low-temperature phase transitions in *hidden sectors* that are independent of Standard Model dynamics have been proposed (107; 108; 109; 106).

4.2.3 Cosmic strings

Cosmic strings are theorized linear (one-dimensional) topological spacetime defects formed as a result of symmetry-breaking cosmological phase transitions in the early Universe (110; 111; 112; 113; 114). Other defects that could form are monopoles and domain walls, although both are fairly confidently ruled out (in the monopole case by "the monopole problem"). Thus cosmic strings are viewed as an exciting potentially-observable consequence of phase transitions in grand unified theories. Such a phase transition would lead to the formation of a *network* of one-dimensional strings permeating the Hubble volume; these strings can "intercommute" when they meet one another (with probability, p), exchanging partners, while string self-interaction can lead to small closed loops being chopped off. The cosmic string network rapidly approaches an attractor scaling regime after formation, where the string correlation length, loop size, and other statistical properties, scales with cosmic time.

The mechanism by which the network reaches the scaling regime is loop decay through GW emission. The cosmic string loops vibrate relativistically under tension (equal to their mass per unit length, μ), leading to the emission of GW signals that causes them to eventually shrink and decay away. These GW signals can sum together incoherently to produce a stochastic GW background that is broadband, with the potential to be constrained by multiple experiments and GW detectors (115; 116; 117; 118). While no models predict a purely power-law spectral shape across the entire GW spectrum, some approximations within the PTA band include $\alpha = -7/6$ and $\gamma = 16/3$ (119; 120) (however, this can vary quite a bit with modeling assumptions (115)). One possible semi-analytic model that is valid within the PTA frequency range is (120)

$$h_c(f) = 1.74 \times 10^{-14} \left(\frac{n_c}{p} \right)^{1/2} \epsilon_{\mathrm{eff}}^{-1/6} \left(\frac{h}{0.7} \right)^{7/6} \left(\frac{G\mu}{10^{-6}} \right)^{1/3} \left(\frac{f}{1\,\mathrm{yr}^{-1}} \right)^{-7/6}$$

$$(4.30)$$

where μ is the cosmic string tension, n_c is the average number of cusps per loop oscillation, p is the intercommutation probability, ϵ_{eff} is a loop length-scale factor, and h is the Hubble constant in units of $100\ \text{km s}^{-1}\text{Mpc}^{-1}$. There is also interest in searching for GW bursts from individual cosmic strings cusps and kinks (121; 122; 123).

4.3 NON-GW SOURCES OF CORRELATED TIMING DELAYS

In an ideal world, the only sources of inter-pulsar correlations in pulsar timing delays would be GWs (modulo the usual statistical noise fluctuations present in any detection scenario). However, there are several sources of timing delays and inter-pulsar correlations that have origins on Earth, in the Solar System, and in the Galaxy, that we must consider.

4.3.1 Clock errors

When first recorded, pulse TOAs are referenced to the local observatory time standard; however these local and national standards are not perfect. Countries distribute national atomic time-scales that are collected and combined by the Bureau International des Poids et Mesures (BIPM)[2], which then issues International Atomic Time (TAI). The latter exists as offsets from the original national atomic time-scales. TAI forms the basis for Coordinated Universal Time (UTC), and ultimately a realization of Terrestrial Time (TT). The relationship between the TAI realization of TT and TAI itself is $\text{TT(TAI)} = \text{TAI} + 32.184$ seconds. Once defined, TAI is never changed, instead being reviewed annually, and with departures from the SI second being corrected to steer TAI into a time-scale realization produced by BIPM (124). This realization is denoted $\text{TT(BIPM\{}year\})$ with $\{year\}$ being the specific numerical year of creation, e.g., 2019. The steering process can lead to a time-scale that is non-stationary, and whose long-timescale stability is difficult to judge. However, at least empirically we observe that recent TT(BIPM) versions differ from TT(TAI) by at least several microseconds since 1994 (125; 126).

Any systematic errors in the realization of TT will be shared by *all* pulsars, since they will all be referenced to the same time standard. This means that every pulsar includes exactly the same clock-error time series, or at least a window of it when baselines are different. The result is a *monopolar* inter-pulsar correlation induced by systematic clock errors, where $\Gamma_{ab} = 1$ for all pulsar pairs (a horizontal line on a correlation–angular-separation diagram) (125; 127; 126). This is often discussed along with *spatially-correlated* processes, or shown on a correlation–angular-separation diagram with the Hellings & Downs curve. While instructive, it is important to note that this is a systematic time-standard error that is common to all pulsars, not some extraterrestrial process that induces monopolar correlations.

---

[2]https://www.bipm.org

4.3.2 Solar-system Ephemeris Errors

As described in Chapter 3, pulsar timing residuals are obtained by subtracting the observed TOAs from a best-fit model. The observed TOAs must be referenced back to the notional emission time at the pulsar, requiring a chain of timing corrections, amongst which are the aforementioned clock corrections. But another important link in this chain is the referencing of observatory-measured TOAs to the equivalent arrival time at the quasi-inertial reference frame of the Solar System Barycenter (SSB). The dominant term in this step is the calculation of the Roemer delay (128), corresponding to the classical light-travel time between Earth and the Sun ($\sim$500 seconds):

$$\Delta_{R\odot} = t^{\mathrm{obs}} - t^{\mathrm{SSB}} = - \left[\vec{r}^{\mathrm{obs}}(t^{\mathrm{obs}}) - \vec{r}^{\mathrm{SSB}}(t^{\mathrm{obs}}) \right] \cdot \hat{p} \qquad (4.31)$$

where $\vec{r}^{\mathrm{obs/SSB}}$ are the coordinate vectors of the observatory and SSB at the observed pulse arrival time, respectively, and $\hat{p}$ is a unit vector pointing in the direction of the pulsar.

Any imperfections in our knowledge of the observatory's or SSB's coordinate vector will induce systematic errors in the barycentering process of Eq. 4.31. Radio observatory positions with respect to the Earth's barycenter are known to within sub-nanosecond precision (129), leaving uncertainties in the SSB position as a possible error source. The resulting systematic timing error can be written as

$$\delta t_{R\odot} = -\delta \vec{x}^{(3)}(t^{\mathrm{obs}}) \cdot \hat{p} \qquad (4.32)$$

where $\vec{x}^{(3)}$ is the Earth barycenter's position in a coordinate frame that has the SSB as the origin. Thus, $\delta \vec{x}^{(3)}$ is the time-dependent systematic error in our Solar System Ephemeris' model of $\vec{x}^{(3)}$. The SSB itself is not an observable position; it is the center of mass of the entire Solar System, whose computation requires accurate masses and orbits of all important dynamical objects, and with quantifiable precision on these values. For pulsar timing analysis, these positions and the SSB calculation rely on published ephemerides, the most prominent of which are the *Development Ephemeris (DE)* series produced by NASA JPL (e.g., 130; 131; 132), and the *Intégration Numérique Planétaire de l'Observatoire de Paris (INPOP)* series produced by IMCCE-Observatoire de Paris (e.g., 133; 134; 135). These ephemeris models require painstaking regression over heterogeneous datasets of different Solar System bodies that stretch back decades, many of which have uncertainties and data quality that is difficult to assess and synthesize with more recent data. Thus formal ephemeris uncertainties and covariances are not considered reliable, and are not published with the best-fit orbits; the latter are used as point estimates in the Roemer delay calculations in pulsar timing.

For reasonable astrophysical models of SMBHBs, the level of induced timing delays is $\sim$100 ns, requiring the position of the SSB to be known to within $\sim\mathcal{O}(100$ m$)$. Until $\sim$2016–2017, these kinds of Solar System ephemeris errors were thought to constitute a sub-dominant source of systematic errors for PTA GW searches. However, several groups found that there were large systematic differences between timing residuals computed under different JPL DE models ranging from DE418 to DE436, whose publication dates differed by more than a decade. This led to an exhaustive analysis by the NANOGrav Collaboration, who found that limits on the SGWB varied significantly when considering ephemerides DE421 to DE436, and the Bayes factor for a common-spectrum process varied by over an order of magnitude (118). This led to the development of a perturbative Bayesian modeling scheme, BAYESEPHEM (118; 136), (which updates the earlier approach of Champion et al. (137)) to model uncertainties on gas giant masses, Jupiter and Saturn's orbital elements, and coordinate frame rotations. This model updates the Roemer delay calculation "in real time" during an MCMC search over the joint SGWB and intrinsic pulsar noise parameter space, contributing additional ephemeris parameters to this global analysis. In the analysis, a baseline DE or INPOP ephemeris is chosen, and the exploration of BAYESEPHEM parameters allows these different ephemerides to "bridge" the systematic differences between them. Therefore current PTA GW searches attempt to constrain the Solar System ephemeris simultaneously with a search for GWs of extragalactic origin. Additional modeling approaches by other PTA groups and the IPTA are in development (138; 139).

The systematic timing error from Solar System ephemeris uncertainties has a cosine dependence on the angle between the SSB error vector and the pulsar's position vector. Thus pulsars on opposite sides of the sky will experience the same magnitude timing error but with an opposite sign. Provided that the SSB error vector is uncorrelated with the pulsar position vectors, Solar System ephemeris errors will induce *dipolar* inter-pulsar correlations, where $\Gamma_{ab} = \cos\theta_{ab}$ (127; 140). This is only an approximation, because of course SSB errors can be modeled deterministically through planetary mass and orbit perturbations (this is how BAYESEPHEM works). Nevertheless, in the regime where there are large number of error sources, it may be effective to treat Solar System ephemeris uncertainties as a dipolar-correlated stochastic process.

4.3.3 Dark Matter

4.3.3.1 *Cold dark Matter Substructure*

Despite being one of the major components of our current concordance cosmology, dark matter is poorly constrained by observations on sub-galactic scales (e.g., 141; 142; 143; 144; 145). This presents problems since well-motivated models of dark matter predict characteristic structure on such small scales, amongst which is the concordance ΛCDM model (Cosmological Constant + Cold Dark Matter) that has a mass function of dark matter haloes that extends down to the free-streaming scale, corresponding to $\sim 10^{-6} M_\odot$ for Weakly Interacting Massive Particle (WIMP) dark matter (146). Constraining dark matter substructure would provide an important validation of CDM and inform limits on the mass of the candidate particle.

High precision pulsar timing could provide a powerful probe of subgalactic dark matter structure, through Doppler and Shapiro effects (e.g., 147; 148; 149; 150). The Doppler effect occurs when dark matter substructure (e.g., a clump, primordial BH, etc.) passes by the Earth or a pulsar, pulling on either and inducing accelerations that change the measured arrival rate of radio pulses. If the substructure passes the Earth then the timing delay will be dipolar-correlated amongst all observed pulsars in an array (similar to the previously discussed Solar System ephemeris systematics), whereas substructure passing close to a pulsar(s) will create uncorrelated timing delays. The Shapiro effect is an integrated timing delay that accumulates as photons propagate through the gravitational potential of substructure along the line of sight between Earth and a given pulsar. As such, it is akin to chromatic timing delays induced by the ionized interstellar medium, in that it builds over the entire path and is uncorrelated between different lines of sight.

The fractional shift to the pulse arrival rate from each effect is given by (151; 152)

$$\left(\frac{\delta\nu}{\nu}\right)_D = \hat{p} \cdot \int \nabla\Phi(\vec{r}, M)dl,$$

$$\left(\frac{\delta\nu}{\nu}\right)_S = -2 \int \vec{v} \cdot \nabla\Phi(\vec{r}, M)dl, \tag{4.33}$$

where $\hat{p}$ is a unit vector in the direction of a pulsar, Φ is the dark matter gravitational potential, M and $\vec{v}$ is the mass and velocity of the dark matter, and l parametrizes the photon propagation path from the pulsar to the Earth.

4.3.3.2 Fuzzy Dark Matter

Fuzzy Dark Matter (FDM) refers to a proposed ultralight axion (essentially a generalization of the QCD axion) as a candidate for dark matter (153; 154). Such an ultralight axion propagating at velocity v would have a macroscopic de Broglie wavelength,

$$\frac{\lambda_{dB}}{2\pi} = \frac{\hbar}{mv} \approx 60\,\text{pc} \left(\frac{10^{-22}\,\text{eV}}{m}\right) \left(\frac{10^{-3}}{v}\right), \tag{4.34}$$

whose wave-like property would effectively suppress power on small scales, thereby resolving some of the problems that cold dark matter models have with predicting more small-scale structure than is observed (155). It is worth noting that FDM can refer to a broader class of ultralight bosons as dark matter candidates.

Tantalisingly, FDM is predicted to produce a measurable influence on pulsar timing observations (156; 157; 158). FDM should couple gravitationally to regular matter in the Milky Way, inducing periodic oscillations in the gravitational potential at frequencies related to twice the mass of the ultralight scalar field:

$$f = \frac{2m}{h} \approx 4.8 \times 10^{-8}\,\text{Hz} \left(\frac{m}{10^{-22}\,\text{eV}}\right). \qquad (4.35)$$

The propagation of photons through this oscillating gravitational potential will lead to shifts in the arrival rate of radio pulses from pulsars. These shifts are not GW-induced; they are entirely due to the time-dependent potential through which the photons propagate. The periodicity in FDM-induced timing delays matches the oscillation of the potential, which happens to lie within the nanohertz PTA sensitivity band for the masses of reasonable FDM candidates. The rms timing delay induced by FDM follows

$$\delta t_{\text{rms}} \approx 0.02\,\text{ns} \left(\frac{m}{10^{-22}\,\text{eV}}\right)^{-3} \left(\frac{\rho_{\text{SF}}}{0.4\,\text{GeV cm}^{-3}}\right), \qquad (4.36)$$

where ρ_{SF} is the local scalar-field dark matter density, and $0.4\,\text{GeV cm}^{-3}$ is the measured local dark matter density (e.g., 159). A full signal model is given in Ref. (160). As the particle mass decreases, the oscillation frequency decreases while the amplitude of induced timing delays increases, thus presenting an exciting opportunity as PTA baselines grow longer (160).

Bibliography

[1] J Kormendy and D Richstone. Inward bound—the search for supermassive black holes in galactic nuclei. *Annual Review of Astronomy and Astrophysics*, 33:581, 1995. 4.1

[2] J Magorrian, S Tremaine, D Richstone, et al. The demography of massive dark objects in Galaxy Centers. *The Astronomical Journal*, 115:2285–2305, June 1998. 4.1

[3] Laura Ferrarese. Beyond the Bulge: A Fundamental Relation between Supermassive Black Holes and Dark Matter Halos. *The Astrophysical Journal*, 578(1):90–97, October 2002. 4.1

[4] Kayhan Gültekin, Douglas O Richstone, Karl Gebhardt, et al. The M-σ and M-L Relations in Galactic Bulges, and Determinations of Their Intrinsic Scatter. *The Astrophysical Journal*, 698(1):198–221, June 2009. 4.1

[5] J Kormendy and LC Ho. Coevolution (Or Not) of Supermassive Black Holes and Host Galaxies. *Annual Review of Astronomy and Astrophysics*, 51:511–653, August 2013. 4.1

[6] Nicholas J McConnell and Chung-Pei Ma. Revisiting the scaling relations of black hole masses and host galaxy properties. *The Astrophysical Journal*, 764(2):184, February 2013. 4.1

[7] Zachary Schutte, Amy E Reines, and Jenny E Greene. The black hole-bulge mass relation including Dwarf Galaxies hosting active galactic nuclei. *The Astrophysical Journal*, 887(2):245, December 2019. 4.1

[8] SDM White and MJ Rees. Core condensation in heavy halos: a two-stage theory for galaxy formation and clustering. *Monthly Notices of the Royal Astronomical Society*, 183:341–358, May 1978. 4.1

[9] Alessandra De Rosa, Cristian Vignali, Tamara Bogdanović, et al. The quest for dual and binary supermassive black holes: A multi-messenger view. *New Astronomy Reviews*, 86:101525, December 2019. 4.1

[10] C Rodriguez, Greg B Taylor, RT Zavala, et al. A compact supermassive binary black hole system. *The Astrophysical Journal*, 646(1):49, 2006. 4.1

[11] Matthew J Graham, S GDjorgovski, Daniel Stern, et al. A systematic search for close supermassive black hole binaries in the Catalina Real-time Transient Survey. *Monthly Notices of the Royal Astronomical Society*, 453(2):1562–1576, October 2015. 4.1

[12] M Charisi, I Bartos, Z Haiman, et al. A population of short-period variable quasars from PTF as supermassive black hole binary candidates. *Monthly Notices of the Royal Astronomical Society*, 463(2):2145–2171, December 2016. 4.1

[13] T Liu, S Gezari, W. Burgett, et al. A systematic search for periodically varying quasars in Pan-STARRS1: An extended baseline test in medium deep survey field MD09. *The Astrophysical Journal*, 833(1):6, December 2016. 4.1

[14] Yu-Ching Chen, Xin Liu, Wei-Ting Liao, et al. Candidate periodically variable quasars from the Dark Energy Survey and the Sloan Digital Sky Survey. *Monthly Notices of the Royal Astronomical Society*, 499(2):2245–2264, December 2020. 4.1

[15] Michael Eracleous, Todd A Boroson, Jules P Halpern, and Jia Liu. A large systematic search for close supermassive binary and rapidly recoiling black holes. *The Astrophysical Journal Supplement*, 201(2):23, August 2012. 4.1

[16] Jessie C Runnoe, Michael Eracleous, Gavin Mathes, et al. A large systematic search for close supermassive binary and rapidly recoiling black holes. II. Continued spectroscopic monitoring and optical flux variability. *The Astrophysical Journal Supplement*, 221(1):7, November 2015. 4.1

[17] Xin Liu, Yue Shen, Fuyan Bian, et al. Constraining Sub-parsec Binary supermassive black holes in quasars with multi-epoch spectroscopy. II. The population with kinematically offset broad balmer emission lines. *The Astrophysical Journal*, 789(2):140, July 2014. 4.1

[18] Jessie C Runnoe, Michael Eracleous, Alison Pennell, et al. A large systematic search for close supermassive binary and rapidly recoiling black holes - III. Radial velocity variations. *Monthly Notices of the Royal Astronomical Society*, 468(2):1683–1702, June 2017. 4.1

[19] Hengxiao Guo, Xin Liu, Yue Shen, et al. Constraining sub-parsec binary supermassive black holes in quasars with multi-epoch spectroscopy - III. Candidates from continued radial velocity tests. *Monthly Notices of the Royal Astronomical Society*, 482(3):3288–3307, January 2019. 4.1

[20] Mohan Rajagopal and Roger W Romani. Ultra–low-frequency gravitational radiation from massive black hole binaries. *The Astrophysical Journal*, 446:543, June 1995. 4.1.1

[21] ES Phinney. A practical theorem on gravitational wave backgrounds. *ArXiv Astrophysics e-prints*, August 2001. 4.1.1

[22] J Stuart B Wyithe and Abraham Loeb. Low-frequency gravitational waves from massive black hole binaries: Predictions for LISA and pulsar timing arrays. *The Astrophysical Journal*, 590(2):691–706, June 2003. 4.1.1

[23] AH Jaffe and DC Backer. Gravitational waves probe the coalescence rate of massive black hole binaries. *The Astrophysical Journal*, 583(2):616–631, February 2003. 4.1.1

[24] Sarah Burke-Spolaor, Stephen R Taylor, Maria Charisi, et al. The astrophysics of nanohertz gravitational waves. *The Astronomy and Astrophysics Review*, 27(1):5, June 2019. 4.1.1, 4.1

[25] PC Peters and J Mathews. Gravitational radiation from point masses in a Keplerian orbit. *Physical Review*, 131:435–440, July 1963. 4.1.1, 4.1.1, 4.1.3, 4.1.3

[26] Philip Carl Peters. Gravitational radiation and the motion of two point masses. *Physical Review*, 136(4B):B1224, 1964. 4.1.1

[27] L Sampson, NJ Cornish, and ST McWilliams. Constraining the solution to the last parsec problem with pulsar timing. *Physical Review D*, 91(8):084055, April 2015. 4.1.2

[28] Siyuan Chen, Hannah Middleton, Alberto Sesana, et al. Probing the assembly history and dynamical evolution of massive black hole binaries with pulsar timing arrays. *Monthly Notices of the Royal Astronomical Society*, 468(1):404–417, June 2017. 4.1.2

[29] Siyuan Chen, Alberto Sesana, and Walter Del Pozzo. Efficient computation of the gravitational wave spectrum emitted by eccentric massive black hole binaries in stellar environments. *Monthly Notices of the Royal Astronomical Society*, 470(2):1738–1749, September 2017. 4.1.2

[30] Siyuan Chen, Alberto Sesana, and Christopher J. Conselice. Constraining astrophysical observables of galaxy and supermassive black hole binary mergers using pulsar timing arrays. *Monthly Notices of the Royal Astronomical Society*, 488(1):401–418, September 2019. 4.1.2

[31] Stephen R Taylor, Joseph Simon, and Laura Sampson. Constraints on the dynamical environments of supermassive black-hole binaries using pulsar-timing arrays. *Physical Review Letters*, 118(18):181102, May 2017. 4.1.2

[32] Luke Zoltan Kelley, Laura Blecha, Lars Hernquist, et al. The gravitational wave Background from massive black hole binaries in Illustris: Spectral features and time to detection with pulsar timing arrays. *arXiv.org*, page arXiv:1702.02180, February 2017. 4.1, 4.1.2.2, 4.1.2.4

[33] LZ Kelley, L Blecha, and L Hernquist. Massive black hole binary mergers in dynamical galactic environments. *Monthly Notices of the Royal Astronomical Society*, 464:3131–3157, January 2017a. 4.1.2.1, 4.1

[34] S Chandrasekhar. Dynamical Friction. I General considerations: the coefficient of dynamical friction. *The Astrophysical Journal*, 97:255, March 1943. 4.1.2.1

[35] F Antonini and D Merritt. Dynamical friction around supermassive black holes. *The Astrophysical Journal*, 745:83, January 2012. 4.1.2.1

[36] D Merritt and M Milosavljević. Massive black hole binary evolution. *Living Reviews in Relativity*, 8, November 2005. 4.1.2.1

[37] Qingjuan Yu. Evolution of massive binary black holes. *Monthly Notices of the Royal Astronomical Society*, 331(4):935–958, April 2002. 4.1.2.1

[38] F Dosopoulou and F Antonini. Dynamical friction and the evolution of supermassive black hole binaries: The final hundred-parsec problem. *The Astrophysical Journal*, 840:31, May 2017. 4.1.2.1

[39] J Binney and S Tremaine. *Galactic Dynamics*. Princeton University Press, Princeton, NJ, 1987. 4.1.2.1

[40] MC Begelman, RD Blandford, and MJ Rees. Massive black hole binaries in active galactic nuclei. *Nature*, 287:307–309, September 1980. 4.1.2.2, 4.1.2.3

[41] J Frank and MJ Rees. Effects of massive central black holes on dense stellar systems. *Monthly Notices of the Royal Astronomical Society*, 176(3):633–647, September 1976. 4.1.2.2

[42] Seppo Mikkola and Mauri J Valtonen. Evolution of binaries in the field of light particles and the problem of two black holes. *Monthly Notices*

of the Royal Astronomical Society (ISSN 0035-8711), 259(1):115–120, November 1992. 4.1.2.2

[43] Gerald D Quinlan. The dynamical evolution of massive black hole binaries I. Hardening in a fixed stellar background. *New Astronomy*, 1(1):35–56, July 1996. 4.1.2.2

[44] Miloš Milosavljević and David Merritt. The Final Parsec Problem. *The astrophysics of gravitational wave sources*, pages 201–210. AIP, 2002. 4.1.2.2

[45] Miloš Milosavljević and David Merritt. Long-term evolution of massive black hole binaries. *The Astrophysical Journal*, 596(2):860–878, October 2003. 4.1.2.2

[46] FM Khan, K Holley-Bockelmann, P Berczik, and A Just. Supermassive black hole binary evolution in axisymmetric galaxies: The final parsec problem is not a problem. *The Astrophysical Journal*, 773:100, August 2013. 4.1.2.2

[47] Eugene Vasiliev and David Merritt. The loss-cone problem in axisymmetric nuclei. *The Astrophysical Journal*, 774(1):87, September 2013. 4.1.2.2

[48] Eugene Vasiliev, Fabio Antonini, and David Merritt. The final-parsec problem in nonspherical galaxies revisited. *The Astrophysical Journal*, 785(2):163, April 2014. 4.1.2.2

[49] Eugene Vasiliev, Fabio Antonini, and David Merritt. The final-parsec problem in the collisionless limit. *The Astrophysical Journal*, 810(1):49, September 2015. 4.1.2.2

[50] Alberto Sesana, Francesco Haardt, and Piero Madau. Interaction of massive black hole binaries with their stellar environment. I. ejection of hypervelocity stars. *The Astrophysical Journal*, 651(1):392–400, November 2006. 4.1.2.2

[51] SR Taylor, J Simon, and L Sampson. Constraints on the dynamical environments of supermassive black-hole binaries using pulsar-timing arrays. *Physical Review Letters*, 118(18):181102, May 2017. 4.1.2.2, 4.1.2.4

[52] Bence Kocsis and Alberto Sesana. Gas-driven massive black hole binaries: signatures in the nHz gravitational wave background. *Monthly Notices of the Royal Astronomical Society*, 411(3):1467–1479, March 2011. 4.1.2.3, 4.1.2.3

[53] PB Ivanov, JCB Papaloizou, and aG Polnarev. The evolution of a supermassive binary caused by an accretion disc. *Monthly Notices of the Royal Astronomical Society*, 307(1):79–90, July 1999. 4.1.2.3, 4.1.2.3

[54] Zoltan Haiman, Bence Kocsis, and Kristen Menou. The population of viscosity- and gravitational wave-driven supermassive black hole binaries among luminous active galactic nuclei. *The Astrophysical Journal*, 700(2):1952–1969, August 2009. 4.1.2.3

[55] Mackenzie SL Moody, Ji-Ming Shi, and James M Stone. hydrodynamic torques in circumbinary accretion disks. *The Astrophysical Journal*, 875(1):66, April 2019. 4.1.2.3

[56] Daniel J D'Orazio and Paul C Duffell. Orbital evolution of equal-mass eccentric binaries due to a gas disk: Eccentric inspirals and circular outspirals. *arXiv e-prints*, page arXiv:2103.09251, March 2021. 4.1.2.3

[57] RM Heath and CJ Nixon. On the orbital evolution of binaries with circumbinary discs. *Astronomy & Astrophysics*, 641:A64, September 2020. 4.1.2.3

[58] Alberto Sesana, Alberto Vecchio, and CN Colacino. The stochastic gravitational-wave background from massive black hole binary systems: implications for observations with Pulsar Timing Arrays. *Monthly Notices of the Royal Astronomical Society*, 390(1):192–209, October 2008. 4.1.2.4

[59] Vicente Rodriguez-Gomez, Shy Genel, Mark Vogelsberger, et al. The merger rate of galaxies in the Illustris Simulation: a comparison with observations and semi-empirical models. *Monthly Notices of the Royal Astronomical Society*, (1):17–64, February 2015. 4.1.2.4

[60] Y Kozai. Secular perturbations of asteroids with high inclination and eccentricity. *The Astronomical Journal*, 67:591, November 1962. 4.1.2.4

[61] ML Lidov. The evolution of orbits of artificial satellites of planets under the action of gravitational perturbations of external bodies. Planet. Space Sci., 9:719–759, October 1962. 4.1.2.4

[62] Junichiro Makino and Toshikazu Ebisuzaki. Triple black holes in the cores of galaxies. *Astrophysical Journal*, 436:607–610, December 1994. 4.1.2.4

[63] Omer Blaes, Man Hoi Lee, and Aristotle Socrates. The Kozai mechanism and the evolution of binary supermassive black holes. *The Astrophysical Journal*, 578(2):775–786, October 2002. 4.1.2.4

[64] Pau Amaro-Seoane, Alberto Sesana, Loren Hoffman, et al. Triplets of supermassive black holes: Astrophysics, gravitational waves and detection. *Monthly Notices of the Royal Astronomical Society*, 402(4):2308–2320, March 2010. 4.1.2.4

[65] Irina Dvorkin and Enrico Barausse. The nightmare scenario: measuring the stochastic gravitational wave background from stalling massive black hole binaries with pulsar timing arrays. *Monthly Notices of the Royal Astronomical Society*, 470(4):4547–4556, 2017. 4.1.2.4

[66] M Bonetti, F Haardt, A Sesana, and E Barausse. Post-Newtonian evolution of massive black hole triplets in galactic nuclei - II. Survey of the parameter space. *Monthly Notices of the Royal Astronomical Society*, 477:3910–3926, July 2018. 4.1.2.4

[67] Matteo Bonetti, Alberto Sesana, Enrico Barausse, and Francesco Haardt. Post-Newtonian evolution of massive black hole triplets in galactic nuclei - III. A robust lower limit to the nHz stochastic background of gravitational waves. *Monthly Notices of the Royal Astronomical Society*, 477(2):2599–2612, June 2018. 4.1.2.4

[68] T Ryu, R Perna, Z Haiman, et al. Interactions between multiple supermassive black holes in galactic nuclei: a solution to the final parsec problem. *Monthly Notices of the Royal Astronomical Society*, 473:3410–3433, January 2018. 4.1.2.4

[69] Eliu Huerta, Sean T Mcwilliams, Jonathan R Gair, and Stephen Taylor. Detection of eccentric supermassive black hole binaries with pulsar timing arrays: Signal-to-noise ratio calculations. *arXiv.org*, (6):063010, April 2015. 4.1.2.4

[70] Motohiro Enoki, M Nagashima, and Masahiro Nagashima. The effect of orbital eccentricity on gravitational wave background radiation from supermassive black hole binaries. *Progress of Theoretical Physics*, 117(2):241–256, February 2007. 4.1.2.4

[71] Vikram Ravi, JSB Wyithe, Ryan M Shannon, et al. Binary supermassive black hole environments diminish the gravitational wave signal in the pulsar timing band. *Monthly Notices of the Royal Astronomical Society*, 442(1):56–68, July 2014. 4.1.2.4

[72] A Rasskazov and D Merritt. Evolution of massive black hole binaries in rotating stellar nuclei: Implications for gravitational wave detection. *Physical Review D*, 95(8):084032, April 2017. 4.1.2.4

[73] SR Taylor, EA Huerta, JR Gair, and ST McWilliams. Detecting eccentric supermassive black hole binaries with pulsar timing arrays: Resolvable source strategies. *The Astrophysical Journal*, 817(1):70, January 2016. 4.1.3, 4.1.3, 4.2, 4.1.3

[74] Leor Barack and Curt Cutler. LISA capture sources: Approximate waveforms, signal-to-noise ratios, and parameter estimation accuracy. *Physical Review D*, 69(8):082005, April 2004. 4.1.3, 4.1.3

[75] Abhimanyu Susobhanan, Achamveedu Gopakumar, George Hobbs, and Stephen R Taylor. Pulsar timing array signals induced by black hole binaries in relativistic eccentric orbits. *Physical Review D*, 101(4):043022, February 2020. 4.1.3

[76] Alberto Sesana and Alberto Vecchio. Measuring the parameters of massive black hole binary systems with pulsar timing array observations of gravitational waves. *Physical Review D*, 81(10):104008, May 2010. 4.1.3

[77] CMF Mingarelli, K Grover, T Sidery, et al. Observing the dynamics of supermassive black hole binaries with pulsar timing arrays. *Physical Review Letters*, 109(8):081104, August 2012. 4.1.3

[78] PC Peters. Gravitational radiation and the motion of two point masses. *Physical Review*, 136:1224–1232, November 1964. 4.1.3

[79] Vincent Corbin and Neil J Cornish. Pulsar timing array observations of massive black hole binaries. *arXiv e-prints*, page arXiv:1008.1782, August 2010. 4.1.3

[80] KJ Lee, N Wex, M Kramer, et al. Gravitational wave astronomy of single sources with a pulsar timing array. *Monthly Notices of the Royal Astronomical Society*, 414(4):3251–3264, July 2011. 4.1.3

[81] JA Ellis. A bayesian analysis pipeline for continuous GW sources in the PTA band. *Classical and Quantum Gravity*, 30(22):224004, November 2013. 4.1.3

[82] PN Payne. Smarr's zero-frequency-limit calculation. *Physical Review D*, 28(8):1894–1897, October 1983. 4.1.4

[83] Demetrios Christodoulou. Nonlinear nature of gravitation and gravitational-wave experiments. *Physical Review Letters*, 67:1486–1489, September 1991. 4.1.4

[84] Luc Blanchet and Thibault Damour. Hereditary effects in gravitational radiation. *Physical Review D*, 46(10):4304–4319, November 1992. 4.1.4

[85] Kip S Thorne. Gravitational-wave bursts with memory: The Christodoulou effect. *Physical Review D*, 45(2):520–524, January 1992. 4.1.4

[86] Marc Favata. Nonlinear gravitational-wave memory from binary black hole mergers. *The Astrophysical Journal*, 696(2):L159–L162, May 2009. 4.1.4

[87] Marc Favata. Post-Newtonian corrections to the gravitational-wave memory for quasicircular, inspiralling compact binaries. *Physical Review D*, 80(2):024002, July 2009. 4.1.4

[88] JM Cordes and FA Jenet. Detecting gravitational wave memory with pulsar timing. *The Astrophysical Journal*, 752(1):54, June 2012. 4.1.4

[89] DR Madison, JM Cordes, and S Chatterjee. Assessing pulsar timing array sensitivity to gravitational wave bursts with memory. *The Astrophysical Journal*, 788(2):141, June 2014. 4.1.4

[90] Rutger van Haasteren and Yuri Levin. Gravitational-wave memory and pulsar timing arrays. *Monthly Notices of the Royal Astronomical Society*, 401(4):2372–2378, February 2010. 4.1.4, 4.1.4

[91] Z Arzoumanian, A Brazier, S Burke-Spolaor, et al. NANOGrav constraints on gravitational wave bursts with memory. *The Astrophysical Journal*, 810(2):150, September 2015. 4.1.4

[92] JB Wang, G Hobbs, W Coles, et al. Searching for gravitational wave memory bursts with the Parkes Pulsar Timing Array. *Monthly Notices of the Royal Astronomical Society*, 446(2):1657–1671, January 2015. 4.1.4

[93] K Aggarwal, Z Arzoumanian, PT Baker, et al. The NANOGrav 11 yr data set: Limits on gravitational wave memory. *The Astrophysical Journal*, 889(1):38, January 2020. 4.1.4

[94] LP Grishchuk. Primordial gravitons and possibility of their observation. *Soviet Journal of Experimental and Theoretical Physics Letters*, 23:293, March 1976. 4.2.1

[95] AA Starobinsky. A new type of isotropic cosmological models without singularity. *Physics Letters B*, 91(1):99–102, March 1980. 4.2.1

[96] AD Linde. A new inflationary universe scenario: A possible solution of the horizon, flatness, homogeneity, isotropy and primordial monopole problems. *Physics Letters B*, 108(6):389–393, February 1982. 4.2.1

[97] M Kamionkowski and ED Kovetz. The Quest for B Modes from inflationary gravitational waves. *Annual Review of Astronomy and Astrophysics*, 54:227–269, September 2016. 4.2.1

[98] Wen Zhao. Constraint on the early Universe by relic gravitational waves: From pulsar timing observations. *Physical Review D*, 83(10):104021, May 2011. 4.2.1

[99] Leonid P Grishchuk. Reviews of topical problems: Relic gravitational waves and cosmology. *Physics Uspekhi*, 48(12):1235–1247, December 2005. 4.2.1

[100] PD Lasky, CMF Mingarelli, TL Smith, et al. Gravitational-Wave Cosmology across 29 Decades in Frequency. *Physical Review X*, 6(1):011035, January 2016. 4.2.1

[101] Mark B Hindmarsh, Marvin Lüben, Johannes Lumma, and Martin Pauly. Phase transitions in the early universe. *arXiv e-prints*, page arXiv:2008.09136, August 2020. 4.2.2

[102] C Caprini, R Durrer, and X Siemens. Detection of gravitational waves from the QCD phase transition with pulsar timing arrays. *Physical Review D*, 82(6):063511, September 2010. 4.2.2

[103] Chiara Caprini and Daniel G Figueroa. Cosmological backgrounds of gravitational waves. *Classical and Quantum Gravity*, 35(16):163001, August 2018. 4.2.2

[104] Chiara Caprini, Mikael Chala, Glauber C Dorsch, et al. Detecting gravitational waves from cosmological phase transitions with LISA: an update. *Journal of Cosmology and Astroparticle Physics*, 2020(3):024, March 2020. 4.2.2

[105] B Allen. The stochastic gravity-wave background: sources and detection. In Jean-Alain Marck and Jean-Pierre Lasota, editors, *Relativistic Gravitation and Gravitational Radiation*, page 373, January 1997. 4.2.2

[106] Zaven Arzoumanian, Paul T Baker, Harsha Blumer, et al. Searching for gravitational waves from cosmological phase transitions with the NANOGrav 12.5-year dataset. *arXiv e-prints*, page arXiv:2104.13930, April 2021. 4.2.2, 4.2.2

[107] Z Chacko, Lawrence J Hall, and Yasunori Nomura. Acceleressence: dark energy from a phase transition at the seesaw scale. *Journal of Cosmology and Astroparticle Physics*, 2004(10):011, October 2004. 4.2.2

[108] Matthew J Strassler and Kathryn M Zurek. Echoes of a hidden valley at hadron colliders. *Physics Letters B*, 651(5-6):374–379, August 2007. 4.2.2

[109] Pedro Schwaller. Gravitational waves from a dark phase transition. *Physical Review Letters*, 115(18):181101, October 2015. 4.2.2

[110] TWB Kibble. Topology of cosmic domains and strings. *Journal of Physics A Mathematical General*, 9(8):1387–1398, August 1976. 4.2.3

[111] Alexander Vilenkin. Gravitational radiation from cosmic strings. *Physics Letters B*, 107(1-2):47–50, December 1981. 4.2.3

[112] A Vilenkin. Cosmic strings and domain walls. Phys. Rep., 121(5):263–315, January 1985. 4.2.3

[113] MB Hindmarsh and TWB Kibble. Cosmic strings. *Reports on Progress in Physics*, 58(5):477–562, May 1995. 4.2.3

[114] A Vilenkin and EPS Shellard. *Cosmic Strings and Other Topological Defects*. 2000. 4.2.3

[115] Sotirios A Sanidas, Richard A Battye, and Benjamin W Stappers. projected constraints on the cosmic (super)string tension with future gravitational wave detection experiments. *The Astrophysical Journal*, 764(1):108, February 2013. 4.2.3

[116] Sotirios A Sanidas, Richard A Battye, and Benjamin W Stappers. Constraints on cosmic string tension imposed by the limit on the stochastic gravitational wave background from the European Pulsar Timing Array. *Physical Review D*, 85(12):122003, June 2012. 4.2.3

[117] L Lentati, SR Taylor, CMF Mingarelli, et al. European Pulsar Timing Array limits on an isotropic stochastic gravitational-wave background. *Monthly Notices of the Royal Astronomical Society*, 453:2576–2598, November 2015. 4.2.3

[118] Z Arzoumanian, PT Baker, A Brazier, et al. The NANOGrav 11 year data set: Pulsar-timing constraints on the stochastic gravitational-wave background. *The Astrophysical Journal*, 859(1):47, May 2018. 4.2.3, 4.3.2

[119] S Ölmez, V Mandic, and X Siemens. Gravitational-wave stochastic background from kinks and cusps on cosmic strings. *Physical Review D*, 81(10):104028, May 2010. 4.2.3

[120] Thibault Damour and Alexander Vilenkin. Gravitational radiation from cosmic (super)strings: Bursts, stochastic background, and observational windows. *Physical Review D*, 71(6):063510, March 2005. 4.2.3

[121] Thibault Damour and Alexander Vilenkin. Gravitational wave bursts from cosmic strings. *Physical Review Letters*, 85(18):3761–3764, October 2000. 4.2.3

[122] Thibault Damour and Alexander Vilenkin. Gravitational wave bursts from cusps and kinks on cosmic strings. *Physical Review D*, 64(6):064008, September 2001. 4.2.3

[123] N Yonemaru, S Kuroyanagi, G Hobbs, et al. Searching for gravitational-wave bursts from cosmic string cusps with the Parkes Pulsar Timing Array. *Monthly Notices of the Royal Astronomical Society*, 501(1):701–712, February 2021. 4.2.3

[124] EF Arias, G Panfilo, and G Petit. Timescales at the BIPM. *Metrologia*, 48(4):S145–S153, August 2011. 4.3.1

[125] G Hobbs, W Coles, RN Manchester, et al. Development of a pulsar-based time-scale. *Monthly Notices of the Royal Astronomical Society*, 427(4):2780–2787, December 2012. 4.3.1

[126] G Hobbs, L Guo, RN Caballero, et al. A pulsar-based time-scale from the International Pulsar Timing Array. *Monthly Notices of the Royal Astronomical Society*, 491(4):5951–5965, February 2020. 4.3.1

[127] C Tiburzi, G Hobbs, M Kerr, et al. A study of spatial correlations in pulsar timing array data. *Monthly Notices of the Royal Astronomical Society*, 455(4):4339–4350, February 2016. 4.3.1, 4.3.2

[128] Olaf Roemer. 1676 a demonstration concerning the motion of light, communicated from paris. *the Journal des Scavans, and here made English. Phil. Trans*, 12:893–894. 4.3.2

[129] RT Edwards, GB Hobbs, and RN Manchester. TEMPO2, a new pulsar timing package - II. The timing model and precision estimates. *Monthly Notices of the Royal Astronomical Society*, 372(4):1549–1574, November 2006. 4.3.2

[130] William M Folkner, James G Williams, and Dale H Boggs. The planetary and lunar ephemeris de 421. *IPN progress report*, 42(178):1–34, 2009. 4.3.2

[131] William M Folkner, James G Williams, Dale H Boggs, et al. The planetary and lunar ephemerides de430 and de431. *Interplanetary Network Progress Report*, 196(1), 2014. 4.3.2

[132] Ryan S Park, William M Folkner, James G Williams, and Dale H Boggs. The jpl planetary and lunar ephemerides de440 and de441. *The Astronomical Journal*, 161(3):105, 2021. 4.3.2

[133] A Fienga, H Manche, J Laskar, et al. INPOP new release: INPOP10e. *arXiv e-prints*, page arXiv:1301.1510, January 2013. 4.3.2

[134] A Fienga, H Manche, J Laskar, et al. INPOP new release: INPOP13b. *arXiv e-prints*, page arXiv:1405.0484, May 2014. 4.3.2

[135] A Fienga, V Viswanathan, P Deram, et al. INPOP new release: INPOP19a. In Christian Bizouard, editor, *Astrometry, Earth Rotation, and Reference Systems in the GAIA era*, pages 293–297, September 2020. 4.3.2

[136] M Vallisneri, SR Taylor, J Simon, et al. Modeling the uncertainties of solar system ephemerides for robust gravitational-wave searches with pulsar-timing arrays. *The Astrophysical Journal*, 893(2):112, April 2020. 4.3.2

[137] DJ Champion, GB Hobbs, RN Manchester, et al. Measuring the mass of solar system planets using pulsar timing. *The Astrophysical Journal*, 720(2):L201–L205, September 2010. 4.3.2

[138] RN Caballero, YJ Guo, KJ Lee, et al. Studying the solar system with the international pulsar timing array. *Monthly Notices of the Royal Astronomical Society*, 481(4):5501–5516, December 2018. 4.3.2

[139] YJ Guo, GY Li, KJ Lee, and RN Caballero. Studying the Solar system dynamics using pulsar timing arrays and the LINIMOSS dynamical model. *Monthly Notices of the Royal Astronomical Society*, 489(4):5573–5581, November 2019. 4.3.2

[140] Elinore Roebber. Ephemeris errors and the gravitational-wave signal: harmonic mode coupling in pulsar timing array searches. *The Astrophysical Journal*, 876(1):55, May 2019. 4.3.2

[141] C Alcock, RA Allsman, DR Alves, et al. The MACHO Project: Microlensing Results from 5.7 Years of Large Magellanic Cloud Observations. *The Astrophysical Journal*, 542(1):281–307, October 2000. 4.3.3.1

[142] P Tisserand, L Le Guillou, C Afonso, et al. Limits on the Macho content of the Galactic Halo from the EROS-2 Survey of the Magellanic Clouds. *Astronomy & Astrophysics*, 469(2):387–404, July 2007. 4.3.3.1

[143] L Wyrzykowski, J Skowron, S Kozłowski, et al. The OGLE view of microlensing towards the Magellanic Clouds - IV. OGLE-III SMC data and final conclusions on MACHOs. *Monthly Notices of the Royal Astronomical Society*, 416(4):2949–2961, October 2011. 4.3.3.1

[144] Kim Griest, Matthew J Lehner, Agnieszka M Cieplak, and Bhuvnesh Jain. Microlensing of kepler stars as a method of detecting primordial black hole dark matter. *Physical Review Letters*, 107(23):231101, December 2011. 4.3.3.1

[145] Steven J Clark, Bhaskar Dutta, Yu Gao, et al. Planck constraint on relic primordial black holes. *Physical Review D*, 95(8):083006, April 2017. 4.3.3.1

[146] Anne M Green, Stefan Hofmann, and Dominik J Schwarz. The first WIMPy halos. *Journal of Cosmology and Astroparticle Physics*, 2005(8):003, August 2005. 4.3.3.1

[147] ER Siegel, MP Hertzberg, and JN Fry. Probing dark matter substructure with pulsar timing. *Monthly Notices of the Royal Astronomical Society*, 382(2):879–885, December 2007. 4.3.3.1

[148] N Seto and A Cooray. Searching for primordial black hole dark matter with pulsar timing arrays. *The Astrophysical Journal*, 659:L33–L36, April 2007. 4.3.3.1

[149] Shant Baghram, Niayesh Afshordi, and Kathryn M Zurek. Prospects for detecting dark matter halo substructure with pulsar timing. *Physical Review D*, 84(4):043511, August 2011. 4.3.3.1

[150] K Kashiyama and N Seto. Enhanced exploration for primordial black holes using pulsar timing arrays. *Monthly Notices of the Royal Astronomical Society*, 426:1369–1373, October 2012. 4.3.3.1

[151] Jeff A Dror, Harikrishnan Ramani, Tanner Trickle, and Kathryn M Zurek. Pulsar timing probes of primordial black holes and subhalos. *Physical Review D*, 100(2):023003, July 2019. 4.3.3.1

[152] Vincent SH Lee, Stephen R Taylor, Tanner Trickle, and Kathryn M Zurek. Bayesian forecasts for dark matter substructure searches with mock pulsar timing data. *arXiv e-prints*, page arXiv:2104.05717, April 2021. 4.3.3.1

[153] Wayne Hu, Rennan Barkana, and Andrei Gruzinov. Fuzzy cold dark matter: the wave properties of ultralight particles. *Physical Review Letters*, 85(6):1158–1161, August 2000. 4.3.3.2

[154] Lam Hui, Jeremiah P Ostriker, Scott Tremaine, and Edward Witten. Ultralight scalars as cosmological dark matter. *Physical Review D*, 95(4):043541, February 2017. 4.3.3.2

[155] Marco Battaglieri, Alberto Belloni, Aaron Chou, et al. US Cosmic Visions: New Ideas in Dark Matter 2017: Community Report. *arXiv e-prints*, page arXiv:1707.04591, July 2017. 4.3.3.2

[156] Andrei Khmelnitsky and Valery Rubakov. Pulsar timing signal from ultralight scalar dark matter. *Journal of Cosmology and Astroparticle Physics*, 2014(2):019, February 2014. 4.3.3.2

[157] NK Porayko and KA Postnov. Constraints on ultralight scalar dark matter from pulsar timing. *Physical Review D*, 90(6):062008, September 2014. 4.3.3.2

[158] Ivan De Martino, Tom Broadhurst, SH Henry Tye, et al. Recognising axionic dark matter by compton and de-broglie scale modulation of pulsar timing. *Physical Review Letters*, 119:221103, November 2017. 4.3.3.2

[159] Jo Bovy and Scott Tremaine. On the local dark matter density. *The Astrophysical Journal*, 756(1):89, September 2012. 4.3.3.2

[160] Nataliya K Porayko, Xingjiang Zhu, Yuri Levin, et al. Parkes pulsar timing array constraints on ultralight scalar-field dark matter. *Physical Review D*, 98(10):102002, November 2018. 4.3.3.2

CHAPTER 5

Data Analysis

5.1 STATISTICAL INFERENCE

Statistical inference is the science of extracting meaning from observation. An astrophysicist can not test a new model without thinking shrewdly about what the observable ramifications will be. Perhaps this new model posits that black holes at the centers of galaxies have a strong influence on the growth and evolution of the galaxy itself. Well … great … but what shall we look at? How do we compute the masses of black holes at the center of galaxies, and what characteristics of the galaxies are measurable in order to test a correlation? In this specific case, my example is a real empirical relationship that exists, showing that there is a strong correlation between central black hole masses and the masses of host-galaxy stellar bulges, bulge luminosity, and bulge velocity dispersion. But even deriving these quantities involves a chain of processes that start at a telescope somewhere, passes through some pipeline to reduce the raw observations to some more digestible form, and involves fitting a model to the reduced observations in order to quote a value with associated precision. Along the way, we must be fastidious in keeping track of sources of systematic and statistical uncertainties so that our final value is a proper reflection of our confidence.

In our case, we are interested in using pulsar-timing data in the form of pulse TOAs (already a reduced form of the initial voltage readings in a receiver) to answer questions like "is there evidence for a stochastic background of gravitational waves?", "what is the most massive binary that could be present in a certain galaxy center as constrained by our data?", and "what kind of radio-frequency dependent noise are the pulses from a certain pulsar experiencing?". There are two main approaches to statistical inference: *frequentist* (or classical) inference, and *Bayesian* inference. The former is predominant in most of the early literature on gravitational waves and even in PTA searches for them, but Bayesian inference now has a foothold as

DOI: 10.1201/9781003240648-5

the flagship approach (as in much of astrophysics research). It has grown steadily in popularity for two key reasons: computing power has made it possible to numerically explore high-dimensional probability distributions, and by incorporating prior model information one can extract meaningful results even from sparse datasets.

Frequentist inference considers the statistical spread of experimental outcomes over a number of trials; hence the use of "frequentist" to denote how probabilities are phrased in this approach. All physical phenomena have true values in the absence of an observation, and all sources of uncertainty and variation are in the data itself. Therefore, the natural question for a frequentist statistician to ask is "how many times should I expect to have measured this data given an assumed model parameter?" So we form a probability distribution for data realizations given a set of model parameters, $p(d|\theta, \mathcal{H})$, where d is data, and θ are parameters of the hypothesis model $\mathcal{H}$. This probability is more often referred to as the *likelihood* function, but note that we talk about the likelihood of parameters given some data, not the likelihood of data – the reader will excuse this dip into pedantry, but there is precedent in the literature (1). Hence, $L(\theta|d) \equiv p(d|\theta)$. By contrast, in Bayesian statistics there is no concept of a true underlying model parameter – there is only the measured data with which we infer some probabilistic distribution of the model parameter. The shift in thinking is subtle, but significant. We are no longer thinking about repeated experimental trials providing a distribution of measured values of a parameter; instead we consider a single measured dataset from which we deduce a probability distribution of a parameter. This Bayesian probability distribution is not a distribution over experimental trials! It is a formal probability distribution that encodes the spread in our belief of what the model parameter is. The terminology for this is the posterior probability $p(\theta|d)$, i.e., the probability of the model parameter given the data. The probabilities at the core of frequentist and Bayesian inference are linked through *Bayes' Theorem*:

$$p(\theta|d) = \frac{p(d|\theta)p(\theta)}{p(d)}, \tag{5.1}$$

where $p(\theta|d)$ is the posterior probability distribution of θ, $p(d|\theta)$ is the likelihood of θ, $p(\theta)$ is the prior probability of θ, and $p(d)$ is a normalization constant that is unimportant in parameter estimation but very important to model selection. It is sometimes called the fully-marginalized likelihood, or evidence, and we will meet it again soon.

In the following I will outline key points in the usage of frequentist and Bayesian statistics. However, the pact of confidence between writer and reader compels me to disclose that I am a card-carrying Bayesian, and this discussion will skew toward that proclivity. Much of the discussion in the sub-section on frequentist inference is inspired by the excellent treatments in Allen & Romano (1999) (2) and Romano & Cornish (2017) (3), while the details of Bayesian inference can be read about in more detail in Jaynes (2003) (1), Gelman *et al.* (2013) (4), and Gregory (2010) (5).

5.2 FREQUENTIST INFERENCE

Classical hypothesis testing assumes that we have repeatable experiments (or trials) with which we can define probabilities as being the fraction of such identical trials in which certain outcomes are realized. A statistic or optimization function is constructed from the data to describe the fitness of a particular model, and although this is often some form of the likelihood function, it need not be. The flexibility in defining the form of statistic is because we are ultimately more interested in its sampling distribution, i.e., the distribution of the statistic over many trials under different hypotheses. These hypothesis distributions for the statistic will determine its detection efficacy. The objectively best statistic is one that meets the *Neyman-Pearson criterion* for maximizing the *detection probability* at a given fixed *false alarm probability*. We will learn exactly what these different terms mean very soon.

5.2.1 Significance

Imagine data being collected from an experiment, with which we compute a detection statistic, $S(d) = S_{\text{obs}}$. As a measure of the fitness of a signal model to the data, we would like to know how often noise fluctuations alone could spuriously produce a statistic as loud as that observed. Thus we need to know the *null hypothesis distribution*, $p(S|\mathcal{H}_{\text{null}})$, i.e. the statistical distribution under the noise-only hypothesis. This distribution could be computed either analytically or through numerical Monte Carlo trials. Either way, it allows us to make a statement about the significance of the measured statistic, denoted by its *p-value* (see also Fig. 5.1 for a graphical description):

$$p = P(S > S_{\text{obs}}|\mathcal{H}_{\text{null}}) = \int_{S_{\text{obs}}}^{\infty} p(S|\mathcal{H}_{\text{null}})dS. \qquad (5.2)$$

There are many abuses of *p*-values, the most egregious of which is interpreting them in terms of support for the signal hypothesis, rather than their true interpretation as arguing against the specific, assumed null hypothesis. Supporting the signal hypothesis requires the statistical distribution under the signal hypothesis, which we next consider.

5.2.2 Type I & II Errors

A pre-defined *p*-value can be assumed in order to set a decision threshold for the statistic. For example, we can adopt a very small *p*-value that maps to a value of the statistic $S_{\text{threshold}}$, above which there is very little chance for noise alone to generate more extreme data. But the chance is not zero! Consider the case where noise does produce a more extreme measured statistic; this would pass our decision threshold, causing us to reject the null hypothesis. This is called a Type I (or more commonly a *false alarm*) error, where

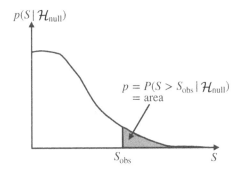

Figure 5.1: The p-value is the probability of a detection statistic exceeding the value measured from some data under the noise-only hypothesis. It reports how often spurious noise fluctuations could give statistic values more extreme than observed. Adapted from Ref. (3).

$S_{\text{obs}} > S_{\text{threshold}}$ under $\mathcal{H}_{\text{null}}$. The converse can also occur, where the statistical distribution of the signal hypothesis allows for the possibility that $S_{\text{obs}} < S_{\text{threshold}}$ under $\mathcal{H}_{\text{signal}}$, causing us to reject the signal hypothesis in a Type II (or *false dismissal*) error. We define the false alarm and false dismissal probabilities as

$$\alpha = P(S > S_{\text{threshold}} | \mathcal{H}_{\text{null}}),$$
$$\beta = P(S < S_{\text{threshold}} | \mathcal{H}_{\text{signal}}), \tag{5.3}$$

where in general the false dismissal probability will depend on the strength of the signal, since this directly influences the spread of $p(S|\mathcal{H}_{\text{signal}})$.

Related to the false dismissal probability is the *detection probability*, $\text{DP} = 1 - \beta$, which measures the fraction of trials in which the signal is correctly identified when truly present (assuming a fixed α that sets the decision threshold). The DP is a quantity that is often tracked in order to determine at what strength we would expect to confidently detect a signal. The strength need not refer simply to the amplitude of the signal; in fact, for PTA GW searches the total observation time and number of pulsars in the array are at least as important factors. Hence we can forecast the detection probability along a number of axes, determining the conditions under which it would surpass a pre-determined threshold of, say, 95% at a fixed false alarm probability of α.

Note that on occasion, people will merely quote the conditions under which the expectation value of the detection statistic S passes the threshold value $S_{\text{threshold}}$. To see why this would be a very weak claim of detection, imagine the statistical distribution of S under the signal hypothesis to be centered on $S_{\text{threshold}}$. Trivially, one can see that the observed value of $S = S_{\text{obs}}$ (which is drawn from $p(S|\mathcal{H}_{\text{signal}})$) only surpasses $S_{\text{threshold}}$ in 50% of the trials. So under these conditions the detection probability is a paltry 50%, no better than a coin flip.

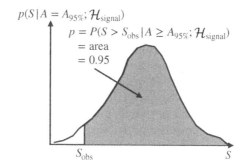

Figure 5.2: The frequentist definition of an upper confidence limit. The distribution of the detection statistic, S is shown under the signal hypothesis when $A = A_{95\%}$. 95% of this distribution lies above the measured value of the statistic, hence we quote $A_{95\%}$ as the frequentist 95% upper limit. Adapted from Ref. (3).

5.2.3 Upper Limits

Without a detection, we can still place constraints that are referenced to upper confidence limits on the amplitude of any signal present. In the frequentist framework, this amounts to saying that if the signal were any louder than this limit, then we would have detected a more extreme value of the statistic than that observed, and with even greater frequency. Thus the upper limit is referenced to a confidence level, and requires knowledge of the statistical distribution of S under the signal hypothesis.

We imagine that in our signal model there is an amplitude parameter, A that we can place upper constraints on. The frequentist 95% upper limit on A is defined as

$$P(S \geq S_{\text{obs}} | A \geq A_{95\%}; \mathcal{H}_{\text{signal}}) = 0.95, \tag{5.4}$$

and shown graphically in Fig. 5.2.

5.3 BAYESIAN INFERENCE

Equation 5.1 states Bayes' Theorem in a form that makes clear the importance of prior knowledge. Indeed, despite the fact that you may hear some people say "but I have not assumed a prior", it is impossible to perform Bayesian inference without assuming a prior, even if it is unconscious. The aforementioned quote usually means that this person has allowed the prior on a parameter to be uniform over an unbounded range, which is often referred to as an *improper* prior because it is not normalizable. We usually attempt to ensure that the posterior probability distribution is data-dominated by employing weakly informative priors, e.g., a wide Gaussian distribution (which is in fact the maximum entropy distribution amongst all real-valued distributions with a specified variance), or a bounded uniform prior in log-space (if the parameter can vary over a large dynamic range). With prior functions that bound the parameters of our model, and a likelihood function that assesses the fitness of some model parameters to the measured data, we can use Bayes' theorem to deduce the posterior probability distribution, $p(\theta|d)$. I will gloss over exactly how that distribution is deduced until later, but it usually involves numerical sampling techniques for which we can ignore the evidence as an unimportant normalization factor. The posterior probability distribution can have arbitrary forms depending on the model and the data; it need not be Gaussian or something else simple. Even if the likelihood is Gaussian, it is only Gaussian in the *data*, not the *parameters*, so one should not expect a simple functional form for the posterior in arbitrary cases.

5.3.1 Parameter Estimation

Consider that our model has more than one parameter, such that $\theta = \{x, y, z\}$. The posterior probability distribution will be multi-dimensional, and we may not even care about all of the parameters being represented. Bayesian inference gives us the power to treat those as *nuisance parameters* that we can *marginalize over*. Marginalization equals integration, so that we collapse our posterior distribution down to a smaller dimensional representation that contains parameters of interest:

$$p(x|d) = \int p(x, y, z|d) \; dydz. \tag{5.5}$$

However, we have not simply fixed the nuisance parameters to some arbitrary values; we have integrated over their entire probabilistic support, and in so doing have propagated all of the uncertainties in those parameters down through to our inference of the parameters of interest.

Parameter estimation in Bayesian inference usually involves quoting quantiles of the one-dimensional marginalized posterior probability distribution of each parameter. Point estimates can be quoted as the mean, median (50% quantile), maximum likelihood, or *maximum-a-posteriori* (MAP)[1] value. Uncertainties are often quoted by defining *credible regions* (named for how Bayesian posterior probabilities encode the spread in our belief of parameters) that are analogous to Gaussian σ's. For example, we may quote the parameter ranges that enclose 68% and 95% of the posterior distribution, which are analogous to Gaussian 1- and 2-σ uncertainties. Credible regions are not unique. There are two ways we can compute an $X\%$ credible region; (i) we integrate (upwards)downwards from $(-)\infty$ until we enclose $X/2\%$ to get the bounding values– this provides an *equal-tailed interval*; or (ii) we imagine lowering a horizontal line from the posterior maximum downwards until we

---

[1] *A posteriori* is a Latin phrase for "from what comes after". You may see it in other textbooks or articles.

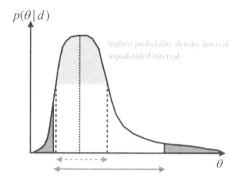

Figure 5.3: Bayesian credible regions encode how well we believe a parameter is constrained within a certain range. However, they are not unique. One can derive an $X\%$ credible interval by integrating the distribution downwards from the maximum-a-posteriori point (this is the *highest probability density interval*), or by integrating inwards from each tail by $(X/2)\%$ (this is the *equal-tailed interval*). Adapted from Ref. (3).

have enclosed $X\%$ of the posterior, where the credible boundaries are where this level intersects the posterior distribution – this is sometimes referred to as the *highest probability density interval*. See Fig. 5.3 to understand how these credible regions are computed in operationally different ways.

It is often useful to investigate covariances amongst parameters in our posterior distribution. To that end we often plot 2-dimensional marginalized posterior distributions. In this case, it makes most sense to compute point estimates and credible regions through the second "level filling" approach mentioned in the previous paragraph. We can even extend our visualization of such covariances by making a "corner" plot (e.g., Fig. 5.4, also known as a "pairwise" or "triangle" plot) with the 2-dimensional distributions of various pairs of parameters occupying positions in a [parameter × parameter] grid, with the relevant 1-dimensional distributions along the diagonal.

5.3.2 Upper Limits

Unlike in frequentist inference, Bayesian inference casts upper limits as a purely parameter estimation issue, by computing one-sided credible regions. To obtain the 95% upper limit on some parameter θ, one simply integrates upward from $-\infty$ (or the relevant prior lower boundary) until 95% of the posterior distribution is enclosed, i.e.,

$$\int_{\theta_{\text{low}}}^{\theta_{95\%}} p(\theta|d)\ d\theta = 0.95, \tag{5.6}$$

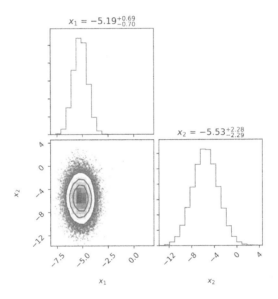

Figure 5.4: An example "corner"/"pairwise"/"triangle" plot, showing marginalized 1-dimensional posterior distributions along the diagonal, and marginalized 2-dimensional distributions of all unique parameter pairs in the lower left corner. Adapted from https://jellis18.github.io/post/2018-01-02-mcmc-part1, and using Ref. (6).

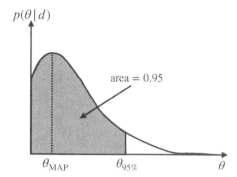

Figure 5.5: Bayesian parameter upper limits are computed by integrating the marginalized parameter posterior distribution upwards until $X\%$ of the posterior mass is enclosed; it is thus treated as a parameter estimation problem. This is markedly different from frequentist upper limits that are based on detection statistics. Adapted from Ref. (3).

where $\theta_{95\%}$ is the 95% upper limit value of θ. In Bayesian inference, an upper limit is a boundary above which there is low credibility for a parameter to have a value.

5.3.3 Model Selection

Parameters belong to a specified hypothesis model. If there are competing parametrized hypotheses to explain a physical phenomenon then we may wish to arbitrate which one the data supports best. There are two main ways that this has so far been performed in PTA searches: (i) *in-sample* techniques that summarize a model's fitness in one number; and (ii) *out-of-sample* techniques that test a model's predictive power.

5.3.3.1 In-sample Model Selection

In Bayesian inference this is performed by computing evidences for each model. As alluded to earlier, the evidence is the fully-marginalized likelihood that appears as a normalization factor in parameter estimation (because in that case we are exploring a fixed model), but must be accounted for in model selection. Basically, it is the average of the prior-weighted likelihood over the entire model parameter space:

$$p(d) := \mathcal{Z} = \int p(d|\theta)p(\theta)d^n\theta. \tag{5.7}$$

This is a challenging integral to evaluate, and I will explain some of the numerical techniques in the next chapter. However with an evidence for each model, one can compute evidence ratios that are more commonly known as

Table 5.1: Loose rule-of-thumb for interpreting a Bayes factor, $\mathcal{B}_{12} = \mathcal{Z}_1/\mathcal{Z}_2$ between models $\mathcal{H}_1$ and $\mathcal{H}_2$ (7).

Bayes factor, $\mathcal{B}$	$\ln(\mathcal{B})$	Strength of evidence
$< 1:1$	< 0	Negative (supports $\mathcal{H}_2$)
$1:1$ to $3:1$	$0-1.1$	Barely worth mentioning
$3:1$ to $10:1$	$1.1-2.3$	Substantial
$10:1$ to $30:1$	$2.3-3.4$	Strong
$30:1$ to $100:1$	$3.4-4.6$	Very strong
$> 100:1$	> 4.6	Decisive

Bayes factors, which themselves may be weighted by prior odds ratios to compute posterior odds ratios:

$$\mathcal{O}_{12} = \frac{P_1}{P_2} \times \frac{\mathcal{Z}_1}{\mathcal{Z}_2}, \quad (5.8)$$

where $\mathcal{O}_{12}$ is the posterior odds of model 1 versus model 2, P_1/P_2 is the prior odds of model 1 versus model 2, and $\mathcal{B}_{12} = \mathcal{Z}_1/\mathcal{Z}_2$ is the Bayes factor of model 1 versus model 2. Prior odds between models are most often set to be equal, giving equivalence between Bayes factor and posterior odds.

But, what does this posterior odds ratio actually mean? Well, it's your betting odds, telling you the strength of belief in one model over another as conditioned on the measured data. The interpretation of Bayes factors and odds ratios are problem specific, but there are some very loose rules of thumb as listed in Table 5.1. However, one must always remember the old acronym *GIGO* from the dawn of the information age, which stands for *"Garbage In, Garbage Out"*. Odds ratios can only rank user-defined models, and if none of them encapsulate a physical model closest to reality then we are just selecting between some bad visions of the phenomenon.

Bayesian evidences ostensibly incorporate the spirit of Occam's razor by penalizing models with excess parameters that may lead to more spread-out likelihood support over the parameter space. However, note that the evidence only penalizes models with parameters that are constrained by the data – no penalty is incurred if a model has many nuisance parameters that are unconstrained, as these will simply integrate against their prior to return one. Priors have a huge effect on Bayesian evidences; naively choosing a prior range that spans orders of magnitude beyond the likelihood support of a dataset can dilute the evidence value, making it more challenging to arbitrate between models.

5.3.3.2 Out-of-Sample Model Selection

This is performed via posterior predictive checks, wherein we test the efficacy of a data-trained model in predicting new data. The posterior predictive density for new data $\tilde{d}$ conditioned on a model $\mathcal{H}$ that has been trained on data, d is

$$p(\tilde{d}|d,\mathcal{H}) = \int p(\tilde{d}|\theta,\mathcal{H})p(\theta|d,\mathcal{H})\, d^n\theta, \qquad (5.9)$$

where we see that *training* on d in this context refers to recovering the posterior probability distribution of model parameters θ, $p(\theta|d,\mathcal{H})$. The other term in the integrand is the probability of new data $\tilde{d}$ given parameters θ, $p(\tilde{d}|\theta)$. One can see the strong analogy with the Bayesian evidence here, except that in this case the prior has been replaced with the posterior of θ as trained on some data d. Given how similar this is in form to the evidence, why would it benefit us to use this posterior predictive density instead? The most important reason is that, despite our best intentions, prior choices can sometimes be somewhat arbitrary in functional choice, and especially in choosing boundaries. As mentioned above, poor prior choices can have a huge impact on evidence values. This is tempered upon using the posterior predictive density, where arbitrary prior choices are replaced by the compact support of a data-trained posterior distribution.

Rather than wait for new data to roll in, we can simply partition the data that we have into training and holdout samples. This is referred to as *cross-validation*, where a common implementation is through k-fold partitioning. The dataset is split into k exclusive subsets, where for each k we obtain the posterior distribution of model parameters as conditioned on the data *not in* k. The posterior predictive density can then be computed for each k subset, and averaged over all subsets. This procedure is repeated for each model under consideration, with differences in the log predictive densities corresponding to ratios between models.

Bibliography

[1] ET Jaynes. *Probability Theory: The Logic of Science.* Cambridge University Press, 2003. 5.1, 5.1

[2] Bruce Allen and Joseph D Romano. Detecting a stochastic background of gravitational radiation: Signal processing strategies and sensitivities. *Physical Review D*, 59(10):102001, May 1999. 5.1

[3] Joseph D Romano and Neil J Cornish. Detection methods for stochastic gravitational-wave backgrounds: a unified treatment. *Living Reviews in Relativity*, 20(1):2, April 2017. 5.1, 5.1, 5.2, 5.3, 5.5

[4] A Gelman, JB Carlin, HS Stern, et al. *Bayesian Data Analysis, Third Edition.* Chapman & Hall/CRC Texts in Statistical Science. Taylor & Francis, 2013. 5.1

[5] Phil Gregory. *Bayesian Logical Data Analysis for the Physical Sciences.* 2010. 5.1

[6] Daniel Foreman-Mackey. corner.py: Scatterplot matrices in python. *The Journal of Open Source Software*, 1(2):24, Jun 2016. 5.4

[7] H Jeffreys. *Theory of probability*. International series of monographs on physics. Clarendon Press, 1983. 5.1

CHAPTER 6

Numerical Bayesian Techniques

All of the techniques mentioned in the previous chapter sound great, but how do we actually implement them? The posterior distributions under study will in general not have a simple functional form, and will have more than just a handful of parameter dimensions. This is important to stress, because there is nothing about the principles of Bayesian inference that demands numerical sampling techniques (like Markov chain Monte Carlo). If the posterior has a simple form, or even only a few dimensions, then you could imagine performing a grid-based mapping of the posterior distribution. But we certainly don't have that; each of our pulsars will have its own ephemeris and noise model, while the GW signal model will have its own set of parameters. The alternative is through random sampling techniques. With a set of random samples, $\{\vec{x}_i\}_N$, drawn from a posterior distribution, $p(\vec{x}|d, \mathcal{H})$, we can perform Monte Carlo integration over arbitrary (multivariate) functions, $f(\vec{x})$:

$$\int f(\vec{x})p(\vec{x}|d, \mathcal{H})d^n x \approx \frac{1}{N}\sum_{i=1}^{N} f(\vec{x}_i). \qquad (6.1)$$

It follows that marginalized posterior probability distributions can easily be obtained by simply binning the random samples in the relevant parameter subset.

So the case is made: we need an efficient path to explore high dimensional distributions, perform random draws, and to perform numerical integrals that correspond to marginalization. Iterative Markov chain Monte Carlo (MCMC) methods are the most often used in PTA GW searches. These make use of the *ergodic theory of Markov chains*, which states that regardless of the initial state of parameter exploration, after a sufficiently large number of iterative steps one will converge towards sampling of the target stationary distribution of interest.

DOI: 10.1201/9781003240648-6

6.1   METROPOLIS ALGORITHMS

Algorithm 1 A typical Metroplis-Hastings algorithm

1: **Initialization** $\vec{x}_{(0)} \sim q(\vec{x})$
2: **for** $i = 1, 2, \ldots$ **do**
3:    Propose:
4:       $\vec{y} \sim q(\vec{y}|\vec{x}_{i-1})$
5:    Acceptance probability:
6:       $\alpha(\vec{y}|\vec{x}_{i-1}) = \min\left\{1, \frac{p(\vec{y})p(d|\vec{y})}{p(\vec{x}_{i-1})p(d|\vec{x}_{i-1})} \times \frac{q(\vec{x}_{i-1}|\vec{y})}{q(\vec{y}|\vec{x}_{i-1})}\right\}$
7:    $u \sim \text{Uniform}(0,1)$
8:    **if** $u < \alpha$ **then**
9:       Accept the proposal: $\vec{x}_i \leftarrow \vec{y}$
10:    **else**
11:       Reject the proposal: $\vec{x}_i \leftarrow \vec{x}_{i-1}$
12:    **end if**
13:    $i = i + 1$
14: **end for**

Metropolis-Hastings sampling is an acceptance-rejection based method of drawing quasi-independent random samples from a target distribution. Moves within parameter space are proposed and then either probabilistically accepted or rejected based on the Metropolis-Hastings ratio (1). A standard Metropolis-Hastings algorithm is shown in Algorithm 1. We draw an initial vector of parameters, $\vec{x}_0$, from the assigned *prior* distribution. At each subsequent iteration, i, a new trial point, $\vec{y}$, is drawn from a *proposal distribution*, $q(\vec{y}|\vec{x}_{i-1})$ and the Metropolis-Hastings ratio is evaluated,

$$\text{MH} = \frac{p(\vec{y})p(d|\vec{y})}{p(\vec{x}_{i-1})p(d|\vec{x}_{i-1})} \times \frac{q(\vec{x}_{i-1}|\vec{y})}{q(\vec{y}|\vec{x}_{i-1})}. \tag{6.2}$$

With this MH value we now need a way to assess whether to accept the proposed point or reject it. We compare MH against a random draw, u, from a uniform distribution, $u \in U[0,1]$. If $u < \text{MH}$ then the move to the new point is accepted and we set $\vec{x}_i = \vec{y}$. If $u > \text{MH}$, the move is rejected and we set $\vec{x}_i = \vec{x}_{i-1}$. There is usually an initial period of exploration where the chain of points is simply roaming around trying to locate regions of high probability in what could be a high-dimensional and potentially multi-modal distribution. This early sequence is usually referred to as the *burn-in*, and discarded from any subsequent usage of the chain. Once the burn-in is over the chain will then be sampling from the target posterior distribution. Over a sufficiently large number of iterations, the chain of visited points corresponds to a collection of (quasi-)independent random draws that, through Monte Carlo integration, can be used for parameter estimation, computing quantiles, and even estimating detection significance.

Including the proposal weighting ratio is crucial in ensuring *detailed balance*, where the transitions in parameters between locations is proportional to the ratio of the posterior densities at those locations. Your exploration will perform best when the proposal distribution is as close as possible to the actual target distribution. To this end, it's possible and completely fine to perform pilot MCMC analyses in order to construct *empirical parameter proposal distributions* for subsequent analyses that improve mixing. Remember that the proposal distribution is not the prior, so you are not double counting here. Just make sure in this case that proposal weightings are correctly implemented in the Metropolis-Hastings ratio. But by far the most common choice for the proposal distribution is that it be symmetric and centered on the current point, in which case we don't need to keep track of the proposal weight factors because that ratio is 1. The simplest choice for such a symmetric proposal distribution is a multi-variate Gaussian distribution with mean equal to the current parameter vector, and a variance tuned to ensure good *mixing*, i.e., the chain is actually exploring the parameter landscape and not simply getting stuck. We'll see more about how to choose this proposal width soon.

The basic visual chain inspections that should be performed for every MCMC problem are shown in Fig. 6.1, corresponding to making *trace-plots* to check chain mixing, histograms to check posterior recovery, and *acceptance rate* monitoring to assess whether the width of the proposal distribution is a problematic factor to the chain's mixing. These are discussed in further detail below.

6.1.1   How Long to Sample?

There are several diagnostics you can use to assess whether you have proceeded through a sufficient number of MCMC iterations. First off, always make *trace-plots*! These are plots of parameter value (or even the log-likelihood value) versus MCMC iteration (e.g., left column of Fig. 6.1). Not only will you be able to judge when to cut off the early burn-in stage, but you will also see whether your chain has settled down to carving out the same region(s) of parameter space. The latter is important, since your chain should be stationary, i.e., statistics computed from one chunk should be equivalent within sampling errors to the statistics of another chunk.

Trace-plots are also an important visual way of judging whether your chain *autocorrelation length* is too long, or whether your proposal acceptance rate is sub-optimal. The former is important because we perform numerical integrals using independent random draws from the target distribution. If the chain is simply wandering around in the same small region then it may be correlated over long timescales, requiring many iterations to fully explore the target distribution. Likewise, if the proposal distribution is attempting

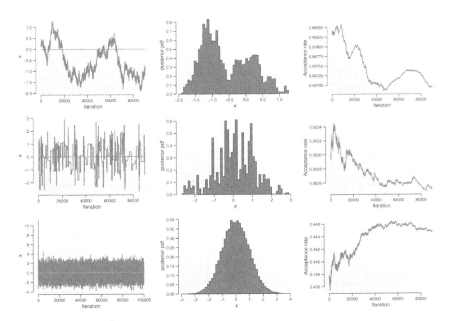

Figure 6.1: Each new MCMC problem should be initially diagnosed with some basic visual checks. Here we search for the mean of a zero-mean unit-variance Gaussian distribution. Each row corresponds to a different MCMC analysis of the same data. The top row has a Gaussian proposal width of $\sigma = 0.01$, the middle row has $\sigma = 500$, while the bottom row has an optimum proposal width of $\sigma = 2.38$. The left panels show the *trace-plot* of the parameter position versus MCMC iteration. The middle panels show the histogram of the MCMC chain (corresponding to the 1-*dimensional marginalized posterior distribution* of the parameter). The right panel shows the *acceptance rate* of proposed points. A short period of *burn-in* is shown in the lower left panel, with the chain starting far from the target region but quickly converging toward it. Adapted from https://jellis18.github.io/post/2018-01-02-mcmc-part1.

large jumps across parameter space to regions of low probability then the chain may not move at all, again leading to high autocorrelation lengths. There are two accepted practices for dealing with autocorrelation lengths larger than a few iterations: (*i*) you can *thin* your chain at the end of sampling by the computed autocorrelation length, ensuring you are using independent random samples; (*ii*) simply run your chain for longer, using brute force to overcome the temporary local clumpiness of correlated samples.

Another important statistic for assessing sampling convergence is the *Gelman-Rubin R statistic* (2). This determines convergence from N independently launched MCMC explorations that initially started from a distribution of points that are "over-dispersed" relative to the variance of the target distribution. When the output from the N chains are indistinguishable, the exploration has converged to the target distribution. The R-statistic compares *in-chain* parameter variance to *inter-chain* parameter variance. The parameter variance from the concatenation of chains is $\hat{\sigma}^2 = [(n-1)W + B]/n$, where W is the mean of in-chain variances, n is the number of MCMC iterations, and B/n is the inter-chain variance. Upon convergence, $\hat{\sigma}^2$ and W should be unbiased. But pre-convergence, $\hat{\sigma}^2$ should under-estimate the variance since parameter mixing has not been sufficient, while W should over-estimate the variance by reflecting the over-dispersal of the initial chain starting points. The R statistic assumes the target distribution is Gaussian, and has the form

$$R = \left\{ [(d+3)\hat{V}]/[(d+1)W] \right\}^{1/2}, \tag{6.3}$$

where $d = 2\hat{V}^2/\mathrm{Var}(\hat{V})$ and $\hat{V} = \hat{\sigma}^2 + B/(nN)$, and $R \to 1$ as the chains converge.

6.1.2 How to Propose New Parameters?

Judging sub-optimal proposal widths is related to the issue of autocorrelation lengths, since too large or too small proposed jumps will lead to poor chain exploration. You will see this in your trace-plot, and most importantly through in-flight *acceptance rate* diagnostics. The ideal acceptance rate of proposed points should be somewhere between $\sim 20-60\%$ (e.g., bottom right panel of Fig. 6.1). Any smaller than this range runs the risk of the chain hardly moving around because the jumps are too big and correspond to regions of low probability (e.g., middle right panel of Fig. 6.1). Any larger than this range and your chain is likely just performing tiny perturbations away from a region of local high probability (e.g., top right panel of Fig. 6.1).

To cut a long story short, you can find the optimum proposal width empirically through many trial MCMCs, or you can let the proposal tune itself. This self-tuning procedure is known as *Adaptive Metropolis MCMC*, which, along with several other alternative proposal schemes, is the subject of the next several sub-sections.

6.1.2.1 Adaptive Metropolis

In Adaptive Metropolis (AM) (3), you use the empirically-estimated parameter covariance matrix to tune the width of the multi-variate Gaussian proposal distribution. Not only this, but tuning is updated during the sampling in order to reach optimal mixing. In practice this means that one uses the entire past history of the chain up until the current point to estimate the parameter covariance matrix[1], scaling this covariance matrix by $\alpha = 2.38^2/n_{\text{dim}}$ to reach the optimal 25% proposal acceptance rate for a target Gaussian distribution. One subtlety here is that by using more than just the most recent point to tune the sampling, our chain is no longer Markovian. This is easily resolved by allowing the chain to pass through a proposal tuning stage using AM, after which the proposal covariance matrix is frozen so that the chain is Markovian then on.

The practical AM strategy is to factorize the chain-estimated parameter covariance matrix, C, using a Cholesky decomposition such that $C = LL^T$. A random draw from this multivariate Gaussian centered on the current point, $\vec{x}_{i-1}$, is given by $\vec{y} = \vec{x}_{i-1} + \alpha^{1/2}L\vec{u}$, where u is an n_{dim} vector of random draws from a zero-mean unit-variance Gaussian. To reduce computational book-keeping time, the parameter covariance matrix can be updated periodically rather than after every single-MCMC iteration.

6.1.2.2 Single Component Adaptive Metropolis

With high-dimensional model parameter spaces, or even target posterior distributions with significant covariances amongst some parameters, the AM method may suffer from low acceptance rates. One method that addresses this is a variant on AM called Single Component Adaptive Metropolis (SCAM) (5). This involves jumping along selected eigenvectors (or principal axes) of the parameter covariance matrix, which is equivalent to jumping in only one *uncorrelated* parameter at a time.

We consider the parameter covariance matrix again, except this time we perform an eigendecomposition on it, $C = D\Lambda D^T$, where D is a unitary matrix with eigenvectors as columns, and $\Lambda = \text{diag}(\sigma_\Lambda^2)$ is a diagonal matrix of eigenvalues. A SCAM jump corresponds to a zero-mean unit-variance jump in a randomly chosen uncorrelated parameter, equivalent to jumping along a vector of correlated parameters. A proposal draw is given by $\vec{y} = \vec{x}_{i-1} + 2.38D_j u_j$, where D_j is a randomly chosen column of D corresponding to the jth eigenvector of C, and $u_j \sim \mathcal{N}(0, (\sigma_\Lambda^j)^2)$.

6.1.2.3 Differential Evolution

Another popular proposal scheme is Differential Evolution (DE) (6), which is a simple genetic algorithm that treats the past history of the chain up until

---

[1]If one uses a recent chunk of the chain rather than the entire past history, then it can be shown that the chain is no longer ergodic (4).

the current point as a *population*. In DE, you choose two random points from the chain's history to construct a difference vector along which the chain can jump. A DE proposal draw is given by $\vec{y} = \vec{x}_{i-1} + \beta(\vec{x}_{r_1} - \vec{x}_{r_2})$, where $\vec{x}_{r_{1,2}}$ are parameter vectors from two randomly chosen points in the past history of the chain, and β is a scaling factor that is usually set to be the same as the AM scaling factor, $\alpha = 2.38^2/n_{\text{dim}}$.

6.1.2.4 The Full Proposal Cocktail

Real world MCMC should use a cocktail of proposal schemes, aimed at ensuring convergence to the target posterior distribution with minimal burn-in, optimal acceptance rate, and as short an autocorrelation length as possible. At each MCMC iteration the proposed parameter location can be drawn according to a weighted list of schemes, involving (i) AM, (ii) SCAM, (iii) DE, (iv) empirical proposal distributions, and finally (iv) draws from the parameter prior distribution. The final prior-draw scheme allows for occasional large departures from regions of high likelihood, ensuring that we are exploring the full parameter landscape well, and avoiding the possibility of getting stuck in local maxima. Really, you can use any reasonable distribution you like to propose points – your only constraint is to ensure that detailed balance is maintained through the relevant proposal weightings in the Metropolis-Hastings ratio.

6.2 GIBBS SAMPLING

Algorithm 2 A typical Gibbs algorithm

1: **Initialization** $x^{(0)} \sim p(x)$
2: **for** $i = 1, 2, \ldots$ **do**
3:     $x_1^{(i)} \sim p(X_1 = x_1 | X_2 = x_2^{(i-1)}, X_3 = x_3^{(i-1)}, \ldots, X_{n_{\text{dim}}} = x_{n_{\text{dim}}}^{(i-1)})$
4:     $x_2^{(i)} \sim p(X_2 = x_2 | X_1 = x_1^{(i)}, X_3 = x_3^{(i-1)}, \ldots, X_{n_{\text{dim}}} = x_{n_{\text{dim}}}^{(i-1)})$
5:     $\vdots$
6:     $x_{n_{\text{dim}}}^{(i)} \sim p(X_{n_{\text{dim}}} = x_{n_{\text{dim}}} | X_1 = x_1^{(i)}, X_2 = x_2^{(i)}, X_3 = x_3^{(i)}, \ldots)$
7:     $i = i + 1$
8: **end for**

Gibbs sampling (7) is an MCMC method that avoids acceptance – rejection techniques, and instead involves sweeping through each parameter (or block of parameters) to draw from their *conditional* probability distributions, with all other parameters fixed to their current values. After sampling for a sufficiently large number of Gibbs steps, the principles of MCMC guarantee that this process of sequential conditional probability draws converges to the joint posterior distribution of the overall model parameter space. A standard Gibbs algorithm is shown in Algorithm 2.

What are the benefits of Gibbs sampling? Well, by drawing directly from the posterior conditionals, the auto-correlation length can be exceptionally small, with minimal burn-in. It's also fast; sequential draws directly from the parameter posterior conditionals means that we are not rejecting any points.

What are the drawbacks of Gibbs sampling? You need to know the form of the conditional probability distributions for each parameter (or parameter blocks), and crucially how to draw samples from it. This can be a non-trivial problem, so typically a lot of effort is placed in manipulating the form of the posterior to find a conditional that is a standard probability distribution. This is where *conjugate priors* become really handy; these are parameter priors for which the the the posterior lies in the same family of distributions as the prior. For example, imagine that we have a Gaussian likelihood function with mean μ and variance σ^2 parameters. The conjugate prior on μ when σ^2 is known (as is assumed in Gibbs when sweeping through each parameter) is a Gaussian distribution, which means that the conditional distribution on μ is simply a Gaussian. Similarly, the conjugate prior on σ^2 when μ is known is an inverse gamma distribution.

All is not lost if a parameter's conjugate prior is not known, or if it is not easy to directly sample from its conditional distribution. You can get inventive by embedding a short Metropolis-Hastings block within the Gibbs algorithm. For example, if there are parameters for which you can not directly draw from the conditional, then your Gibbs step for that parameter could be a short Metropolis-Hastings MCMC run. The goal is to run this until you have drawn a single quasi-independent random sample from the parameter posterior conditional distribution, with all other parameters fixed. You can then proceed through the remainder of your Gibbs steps as normal.

6.3 EVIDENCE EVALUATION AND MODEL SELECTION

As mentioned earlier, evaluating the Bayesian evidence is fraught with issues of ensuring adequate sampling of the prior volume, possible multi-modalities, and issues of computational power. There are some practical strategies with which to tackle it though, which I now briefly discuss. In what follows I will change notation such that model parameters are now labeled θ, which can be a vector.

6.3.1 Harmonic Mean Estimator

It is possible to use an MCMC chain to directly compute the Bayesian evidence, $\mathcal{Z}$ (8). Remember that Bayes' theorem can be phrased as

$$\frac{p(\theta)}{\mathcal{Z}} = \frac{p(\theta|d)}{p(d|\theta)}. \tag{6.4}$$

Integrating both sides over θ gives

$$\frac{1}{\mathcal{Z}} = \int \frac{p(\theta|d)}{p(d|\theta)} \, d\theta, \tag{6.5}$$

since the prior is normalized to one, and $\mathcal{Z}$ is a constant. This means that with a sufficiently converged MCMC exploration of a model parameter space, we can use the MCMC samples to perform Monte Carlo integration to deliver the harmonic mean estimator of the evidence:

$$\mathcal{Z} = \left\langle \frac{1}{p(d|\theta_i)} \right\rangle^{-1}_{i=1,\dots,N}, \tag{6.6}$$

where the angled brackets denote the expectation over the MCMC samples, $\{\theta_1, \dots, \theta_N\}$. This estimator is notoriously unreliable though, and is almost never employed for precision evidence calculation or model selection.

6.3.2 Information Criterion Proxies

There are several computationally-cheap proxies for the evidence that are based on Taylor expansions of the log likelihood function around its maximum.

6.3.2.1 Bayesian Information Criterion

Also known as the Schwarz Information Criterion (9), the Bayesian Information Criterion (BIC) is derived through an expansion of the log-likelihood function up to quadratic order around its maximum, followed by integrating over the model parameters to compute the evidence. In the limit of large numbers of observations, N, that exceed the number of model parameters, k, the BIC of model $\mathcal{H}$ can be written as

$$\mathrm{BIC}(\mathcal{H}|d) \equiv -2\ln \mathcal{Z}_{\mathcal{H}} \approx k\ln N - 2\ln p(d|\hat{\theta}_{\mathrm{MLE}}) \tag{6.7}$$

where the second term after the approximation equality is the log-likelihood evaluated at the maximum likelihood parameters, $\hat{\theta}_{\mathrm{MLE}}$.

For a Gaussian likelihood, the BIC is linearly related to the χ^2 value, but incorporates an additional level of parsimony to penalize model complexity based on the number of parameters. In model selection, the goal is to achieve the lowest BIC value possible among the ensemble of tested models. The rule of thumb with $\Delta\mathrm{BIC}$ values is that $0-2$ is not worth mentioning, $2-6$ is positive, $6-10$ is strong, and >10 is very strong evidence against the model with the higher BIC (10).

6.3.2.2 Akaike Information Criterion

The Akaike Information Criterion (AIC) is another information criterion based on similar approximations as the BIC, but penalizes model complexity less severely (11):

$$\mathrm{AIC}(\mathcal{H}|d) = 2k - 2\ln p(d|\hat{\theta}_{\mathrm{MLE}}). \tag{6.8}$$

In fact, there is an additional form of the AIC that is corrected for small sample sizes (12):

$$\text{AICc}(\mathcal{H}|d) = \text{AIC}(\mathcal{H}|d) + \frac{2k(k+1)}{N-k-1} = 2k + \frac{2k(k+1)}{N-k-1} - 2\ln p(d|\hat{\theta}_{\text{MLE}}).$$
(6.9)

The model selection aim with AIC is the same as with BIC, where we try to find the model that minimizes the AIC value. However, note that both the AIC and BIC penalize model complexity based on the number of parameters, regardless of whether the parameter is constrained or not. This is different from the full Bayesian evidence, where parameters are only penalized if they are constrained by the data (but are nevertheless superfluous); parameters that are unconstrained by the data will simply produce a constant likelihood, integrating against their prior and canceling out in a ratio of model evidences.

6.3.3 Thermodynamic Integration

This evidence calculation technique employs *parallel tempering*, which is a method of launching many MCMC chains of varying "temperature", T, designed to aggressively search parameter space and avoid trapping of chains in local likelihood maxima (13; 14). A chain's temperature denotes the degree to which the likelihood contrast is smoothed, where a $T = \infty$ chain is essentially exploring the prior space. Each chain has a different target distribution, $p(\theta|d, \beta) = p(\theta)p(d|\theta)^\beta$, where $\beta = 1/T \in [0, 1]$. Higher temperature chains are more capable of easily exploring regions far from the likelihood maximum. A multi-temperature Hastings step is used to ensure inter-chain mixing and rapid localization of the global maximum, where the multi-T Hastings ratio is

$$H_{i\to j} = \frac{p(d|\theta_i, \beta_j)p(d|\theta_j, \beta_i)}{p(d|\theta_i, \beta_i)p(d|\theta_j, \beta_j)}.$$
(6.10)

The evidence for a chain with inverse temperature β is (see, e.g., 15, and references therein)

$$\mathcal{Z}_\beta = \int p(\theta)p(d|\theta)^\beta \, d\theta,$$
(6.11)

such that

$$\begin{aligned}
\ln \mathcal{Z} &= \int_0^1 \frac{\partial \ln \mathcal{Z}}{\partial \beta} \, d\beta \\
&= \int_0^1 d\beta \int \frac{p(\theta)p(d|\theta)^\beta}{\mathcal{Z}_\beta} \ln p(d|\theta) \, d\theta \\
&= \int_0^1 \langle \ln p(d|\theta) \rangle_\beta \, d\beta,
\end{aligned}$$
(6.12)

where angled brackets denote the expectation over the MCMC chain with inverse temperature β. This is an exact method for evidence evaluation, in the sense that the limitations on the evidence precision are through practical considerations like sampling efficiency, and the design of the temperature spacings (sometimes called the *temperature ladder*). The choice of temperature-ladder spacing and maximum temperature is problem- and dimension-dependent; for an example of the ladder construction for individual SMBBH searches with PTAs, see Ref. (16; 17), and for more general GW-search diagnostics, see Ref. (18).

6.3.4 Nested sampling

The most ubiquitous implementation of this Monte Carlo technique (originally proposed by Skilling (19)) is the MULTINEST algorithm (see Ref. (20; 21; 22)). We discuss this algorithm in the following, but note that there are many other variants (e.g., 23; 24; 25; 26). A model's parameter space is populated with "live" points drawn from the prior. These points climb through nested contours of increasing likelihood, where at each iteration the points are ranked by likelihood such that the lowest ranked point is replaced by a higher likelihood substitute. The latter step poses the biggest difficulty– drawing new points from the prior volume provides a steadily decreasing acceptance rate, since at later iterations the live-set occupies a smaller volume of the prior space. MULTINEST uses an ellipsoidal rejection-sampling technique, where the current live-set is enclosed by (possibly overlapping) ellipsoids, and a new point drawn uniformly from the enclosed region. This technique successfully copes with multimodal distributions and parameter spaces with strong, curving degeneracies.

The multi-dimensional evidence integral is calculated by transforming to a one-dimensional integral that is easily numerically evaluated. The prior volume, X is defined as

$$dX = p(\theta)d^n\theta, \tag{6.13}$$

such that

$$X(\lambda) = \int_{\mathcal{L}(\theta)>\lambda} p(\theta)d^n\theta, \tag{6.14}$$

where the integral extends over the region of the n-dimensional parameter space contained within the iso-likelihood contour $\mathcal{L}(\theta) = \lambda$. Thus the evidence integral can be written as

$$\mathcal{Z} = \int p(d|\theta)p(\theta)\, d^n\theta = \int_0^1 \mathcal{L}(X)\, dX, \tag{6.15}$$

where $\mathcal{L}(X)$ is a monotonically decreasing function of X. Ordering the X values allows the evidence to be approximated numerically using the trapezium rule

$$\mathcal{Z} = \sum_{i=1}^M \mathcal{L}_i w_i, \tag{6.16}$$

where the weights, w_i, are given by $w_i = (X_{i-1} - X_{i+1})/2$.

Although designed for evidence evaluation, the final live-set and all discarded points can be collected and assigned probability weights to compute the posterior probability of each point. These points can be used to deduce posterior integrals as in other MCMC methods.

6.3.5 Savage-Dickey density ratio

We consider the case of nested models, $\mathcal{H}_1$, $\mathcal{H}_2$ that share many common parameters, θ. However one of the models, $\mathcal{H}_1$ includes a signal that can be switched on and off by some parameter, A. Identifying this parameter (or corner of parameter space) that nulls the signal is the key to implementing this Savage-Dickey approach (27), such that

$$p(d|A = 0, \theta; \mathcal{H}_1) = p(d|\theta; \mathcal{H}_2). \qquad (6.17)$$

Let's work out the posterior density for $A = 0$:

$$p(A = 0|d; \mathcal{H}_1) = \int p(A = 0, \theta|d; \mathcal{H}_1) \, d^n\theta \qquad (6.18)$$

$$= \int \frac{p(d|A = 0, \theta; \mathcal{H}_1)p(A = 0)p(\theta)}{p(d|\mathcal{H}_1)} \, d^n\theta$$

$$= \int \frac{p(d|\theta; \mathcal{H}_2)p(A = 0)p(\theta)}{p(d|\mathcal{H}_1)} \, d^n\theta$$

$$= \frac{p(A = 0)}{p(d|\mathcal{H}_1)} \int p(d|\theta; \mathcal{H}_2)p(\theta) \, d^n\theta$$

$$= \frac{p(d|\mathcal{H}_2)}{p(d|\mathcal{H}_1)} p(A = 0), \qquad (6.19)$$

where we have used Bayes' Theorem on the second line, and Eq. 6.17 on the third line. We can see immediately that the Bayes factor between model 1 and model 2 is

$$\mathcal{B}_{12} = \frac{p(d|\mathcal{H}_1)}{p(d|\mathcal{H}_2)} = \frac{p(A = 0)}{p(A = 0|d; \mathcal{H}_1)}, \qquad (6.20)$$

corresponding to the ratio of the prior to marginal posterior densities of $A = 0$. If the data is informative of a non-zero signal being present then the posterior support at $A = 0$ will be less than the prior support, such that $\mathcal{B}_{12} > 1$, as desired.

The Savage-Dickey density ratio has been used extensively in PTA searches for GWs, where the amplitude of a fixed spectral-index GWB is the relevant on/off signal switch. For practical purposes, the position of zero signal is the lower sampling limit of the GWB strain amplitude, $\log_{10} A_{\text{low}} = -18$. This limit is chosen to make the lowest GWB signal level much smaller than the typical intrinsic noise levels in the pulsars. The recovered Bayes factor is between a model with intrinsic pulsar noise plus a GWB versus a model with noise alone. This does not immediately give a measure of the significance of GW-induced inter-pulsar correlations, which is the most sought-after statistic for PTA searches.

6.3.6 Product Space Sampling

In this approach, we treat model selection as a parameter estimation problem (28; 29; 30; 31). We define a hypermodel, $\mathcal{H}_*$, whose parameter space is the concatenation of all sub-model spaces under consideration, along with an additional model tag, n. This model tag is discrete, but in the process of sampling we can simply apply appropriate model behaviour within some bounded region of a continuously sampled variable.

At a given iteration in the sampling process we evaluate the model tag's position that indicates the "active" sub-model to be used for the likelihood evaluation. The hypermodel parameter space, θ_*, is filtered to find the relevant parameters of the active sub-model, which are then passed to the likelihood function. The parameters of the inactive sub-models do not contribute to, and are not constrained by, the active likelihood function. As sampling proceeds, the model tag will vary between all sub-models such that the relative fraction of iterations spent in each sub-model provides an estimate of the posterior odds ratio.

Let's write the marginalized posterior distribution of the model tag from the output of MCMC sampling:

$$p(n|d; \mathcal{H}_*) = \int p(\theta_*, n|d; \mathcal{H}_*)d\theta$$

$$= \frac{1}{\mathcal{Z}_*} \int p(d|\theta_*, n; \mathcal{H}_*)p(\theta_*, n|\mathcal{H}_*)d\theta, \qquad (6.21)$$

where $\mathcal{Z}_*$ is the hypermodel evidence. For a given n the hypermodel parameter space is partitioned into active, θ_n, and inactive, $\theta_{\bar{n}}$, parameters, $\theta_* = \{\theta_n, \theta_{\bar{n}}\}$, where the likelihood $p(d|\theta_*, n; \mathcal{H}_*)$ is independent of the inactive parameters. The prior term can then be factorized:

$$p(\theta_*, n|\mathcal{H}_*) = p(\theta_n|\mathcal{H}_n)p(\theta_{\bar{n}}|\mathcal{H}_{\bar{n}})p(n|\mathcal{H}_*), \qquad (6.22)$$

such that

$$p(n|d, \mathcal{H}_*) = \frac{p(n|\mathcal{H}_*)}{\mathcal{Z}_*} \int p(d|\theta_n; \mathcal{H}_n)p(\theta_n|\mathcal{H}_n)d\theta_n$$

$$= \frac{p(n|\mathcal{H}_*)}{\mathcal{Z}_*} \mathcal{Z}_n, \qquad (6.23)$$

where $\mathcal{Z}_n$ is the evidence for sub-model n. We have already marginalized over inactive parameters since they only appear in the prior term $p(\theta_{\bar{n}}|\mathcal{H}_{\bar{n}})$, which integrates to one. Thus the posterior odds ratio between two models is given by:

$$\mathcal{O}_{12} = \frac{p(n_1|\mathcal{H}_*)\mathcal{Z}_1}{p(n_2|\mathcal{H}_*)\mathcal{Z}_2} = \frac{p(n_1|d;\mathcal{H}_*)}{p(n_2|d;\mathcal{H}_*)}, \tag{6.24}$$

where the hypermodel evidence cancels in this ratio of the two sub-models.

We can refine this method further by designing an iterative scheme, where pilot runs provide an initial estimate of $\tilde{\mathcal{O}}_{12}$, which is then followed by focused runs for an improved estimate. In a focused run, we weight model 1 by $1/(1 + \tilde{\mathcal{O}}_{12})$ and model 2 by $\tilde{\mathcal{O}}_{12}/(1 + \tilde{\mathcal{O}}_{12})$. This improves mixing and exploration by reducing the posterior contrast across the model landscape. The posterior odds ratio from the focused run is then re-weighted by the pilot run estimate to provide a more accurate value of $\mathcal{O}_{12}$. The product-space estimator (sometimes called a *hypermodel analysis* in PTA circles) of the posterior odds ratio is simple to implement, applicable to high dimensional parameter spaces, and allows direct model comparison.

Product space sampling is being used within PTA data analysis to deduce bespoke noise models for the time-series of each pulsar, and to assess the odds of our data showing inter-pulsar correlations matching the Hellings & Downs curve. The implementation can be found in `enterprise_extensions`[2].

Bibliography

[1] W Keith Hastings. Monte carlo sampling methods using markov chains and their applications. 1970. 6.1

[2] Inference from iterative simulation using multiple sequences. *Statistical Science*, 7(4):457–472, 1992. 6.1.1

[3] Heikki Haario, Eero Saksman, Johanna Tamminen, et al. An adaptive metropolis algorithm. *Bernoulli*, 7(2):223–242, 2001. 6.1.2.1

[4] Heikki Haario, Eero Saksman, and Johanna Tamminen. Adaptive proposal distribution for random walk metropolis algorithm. *Computational Statistics*, 14(3):375–396, 1999. 1

[5] Heikki Haario, Eero Saksman, and Johanna Tamminen. Componentwise adaptation for high dimensional mcmc. *Computational Statistics*, 20(2):265–273, 2005. 6.1.2.2

[6] Cajo JF Ter Braak. A markov chain monte carlo version of the genetic algorithm differential evolution: easy bayesian computing for real parameter spaces. *Statistics and Computing*, 16(3):239–249, 2006. 6.1.2.3

[7] S Geman and D Geman, Stochastic relaxation, Gibbs distributions, and the Bayesian restoration of images, *IEEE Transactions on pattern analysis and machine intelligence*, PAMI-6(6):721–741, 1984, doi: 10.1109/TPAMI.1984.4767596. 6.2

---

[2]https://github.com/nanograv/enterprise_extensions

[8] Michael A Newton and Adrian E Raftery. Approximate bayesian inference with the weighted likelihood bootstrap. *Journal of the Royal Statistical Society. Series B (Methodological)*, 56(1):3–48, 1994. 6.3.1

[9] Gideon Schwarz et al. Estimating the dimension of a model. *Annals of statistics*, 6(2):461–464, 1978. 6.3.2.1

[10] Robert E Kass and Adrian E Raftery. Bayes factors. *Journal of the american statistical association*, 90(430):773–795, 1995. 6.3.2.1

[11] Hirotugu Akaike. A new look at the statistical model identification. *IEEE transactions on automatic control*, 19(6):716–723, 1974. 6.3.2.2

[12] Andrew R Liddle. Information criteria for astrophysical model selection. *Monthly Notices of the Royal Astronomical Society: Letters*, 377(1):L74–L78, 2007. 6.3.2.2

[13] Andrew Gelman and Xiao-Li Meng. Simulating normalizing constants: From importance sampling to bridge sampling to path sampling. *Statistical science*, pages 163–185, 1998. 6.3.3

[14] Yosihiko Ogata. A monte carlo method for high dimensional integration. *Numerische Mathematik*, 55(2):137–157, 1989. 6.3.3

[15] Tyson B Littenberg and Neil J Cornish. Bayesian approach to the detection problem in gravitational wave astronomy. *Physical Review D*, 80(6):063007, September 2009. 6.3.3

[16] JA Ellis. A Bayesian analysis pipeline for continuous GW sources in the PTA band. *Classical and Quantum Gravity*, 30(22):224004, November 2013. 6.3.3

[17] Z Arzoumanian, A Brazier, S Burke-Spolaor, et al. Gravitational Waves from Individual Supermassive Black Hole Binaries in Circular Orbits: Limits from the North American Nanohertz Observatory for Gravitational Waves. *The Astrophysical Journal*, 794(2):141, October 2014. 6.3.3

[18] Tyson B Littenberg and Neil J Cornish. Separating gravitational wave signals from instrument artifacts. *Physical Review D*, 82(10):103007, November 2010. 6.3.3

[19] John Skilling. Nested sampling. In *AIP Conference Proceedings*, volume 735, pages 395–405. American Institute of Physics, 2004. 6.3.4

[20] F Feroz and MP Hobson. Multimodal nested sampling: an efficient and robust alternative to Markov Chain Monte Carlo methods for astronomical data analyses. *Monthly Notices of the Royal Astronomical Society*, 384(2):449–463, February 2008. 6.3.4

[21] F Feroz, MP Hobson, and M Bridges. MULTINEST: an efficient and robust Bayesian inference tool for cosmology and particle physics. *Monthly Notices of the Royal Astronomical Society*, 398(4):1601–1614, October 2009. 6.3.4

[22] Farhan Feroz, Michael P Hobson, Ewan Cameron, and Anthony N. Pettitt. Importance Nested Sampling and the MultiNest Algorithm. *The Open Journal of Astrophysics*, 2(1):10, November 2019. 6.3.4

[23] WJ Handley, MP Hobson, and AN Lasenby. POLYCHORD: nextgeneration nested sampling. *Monthly Notices of the Royal Astronomical Society*, 453(4):4384–4398, November 2015. 6.3.4

[24] Matthew Pitkin, Maximiliano Isi, John Veitch, and Graham Woan. A nested sampling code for targeted searches for continuous gravitational waves from pulsars. *arXiv e-prints*, page arXiv:1705.08978, May 2017. 6.3.4

[25] Rory Smith, Gregory Ashton, Avi Vajpeyi, and Colm Talbot. Massively parallel Bayesian inference for transient gravitational-wave astronomy. *arXiv e-prints*, page arXiv:1909.11873, September 2019. 6.3.4

[26] Joshua S Speagle. DYNESTY: a dynamic nested sampling package for estimating Bayesian posteriors and evidences. *Monthly Notices of the Royal Astronomical Society*, 493(3):3132–3158, April 2020. 6.3.4

[27] James M Dickey. The weighted likelihood ratio, linear hypotheses on normal location parameters. *The Annals of Mathematical Statistics*, 42(1):204–223, 1971. 6.3.5

[28] Bradley P Carlin and Siddhartha Chib. Bayesian model choice via markov chain monte carlo methods. *Journal of the Royal Statistical Society: Series B (Methodological)*, 57(3):473–484, 1995. 6.3.6

[29] Simon J Godsill. On the relationship between markov chain monte carlo methods for model uncertainty. *Journal of computational and graphical statistics*, 10(2):230–248, 2001. 6.3.6

[30] Z Arzoumanian, PT Baker, A Brazier, et al. The NANOGrav 11 Year Data Set: Pulsar-timing Constraints on the Stochastic Gravitational-wave Background. *The Astrophysical Journal*, 859(1):47, May 2018. 6.3.6

[31] Stephen R Taylor, Rutger van Haasteren, and Alberto Sesana. From bright binaries to bumpy backgrounds: Mapping realistic gravitational wave skies with pulsar-timing arrays. *Physical Review D*, 102(8):084039, October 2020. 6.3.6

CHAPTER 7

The PTA Likelihood

Constructing the PTA likelihood begins with an understanding of the dominant influences on the pulse TOAs. To leading order, this is of course the deterministic pulsar timing ephemeris, depending on the pulse period, spindown rate, astrometric effects, radio-frequency dependent delays due to propagation through the ionized interstellar medium, etc. Pulsar-timing astronomers diligently construct these timing ephemerides over repeated observations, accruing more evidence for marginal effects that may be related to binary orbital effects, such as e.g., Shapiro delay. The result of such painstaking work over years and years is a best-fit timing ephemeris with respect to leading-order noise processes in the TOAs, such as radiometer noise (really just pulse shape template-fitting errors), low-frequency/long-timescale noise, etc.

The search for GWs is in the *timing residuals*, literally the remnant time-series upon subtracting the best-fit timing ephemeris from the raw observed TOAs. Anything left after this subtraction should merely be noise and GWs. "But wait!", I hear you holler, with righteous indignation; "What if the original timing-ephemeris fit accidentally removed some of the GW signal? Won't this bias or hinder the search?" This is of course true, so let's take a dive into the pulsar-timing data model.

Throughout the following it will be convenient to consider both sampling frequencies of a pulsars' time-series data, f, and the radio frequencies at which the TOAs are observed, ν. Unless otherwise stated, "frequency" will be taken to mean the sampling frequency of the time-series, which is often synonymous with the frequency of GWs being probed. A final note that a detailed study of the computational evaluation time of each part of the PTA likelihood will not be given in the following; readers interested in that should see the comprehensive overview given in Ref. (1).

DOI: 10.1201/9781003240648-7

7.1 THE PULSAR-TIMING DATA MODEL

A pulsar's TOAs can be written as a sum of deterministic and stochastic parts

$$\vec{t}_{\text{TOA}} = \vec{t}_{\text{det}} + \vec{t}_{\text{stoch}}, \tag{7.1}$$

where $\vec{t}_{\text{TOA}}$ is the vector of pulse arrival times of length N, and $\vec{t}_{\text{det/stoch}}$ are the corresponding deterministic and stochastic components. We model all stochastic contributions as Gaussian random processes; there are important reasons for this that will be seen in detail later, but essentially this means that all statistical properties can be encapsulated in the first and second moments of the stochastic time-series, allowing us to use some of the lovely properties of Gaussian distributions and their integrals when formulating the PTA likelihood[1].

7.1.1 Timing Ephemeris

Assuming that we have some estimate of the m timing-ephemeris parameters from the Herculean efforts of pulsar-timing astronomers, $\vec{\beta}_0$, we can form the pulsar-timing residuals, $\delta\vec{t}$

$$\delta\vec{t} \equiv \vec{t}_{\text{TOA}} - \vec{t}_{\text{det}}(\vec{\beta}_0). \tag{7.2}$$

Ideally, we would like to vary the timing-ephemeris parameters at the same time as we search for GWs, thus ensuring that we don't erroneously remove some of the signal through this initial fitting procedure. I'll explain later how we can do this explicitly, but for now we're going to try to be smart about this. Realistically, any GW-induced perturbation to the TOAs will be miniscule, swamped by the determinstic effects and noise processes. Therefore we can safely assume that the difference between the initial best-fit parameters $\vec{\beta}_0$ and those we would get from the full analysis that includes GWs, $\vec{\beta}_{\text{full}}$, is very, very small. We can then linearize the deterministic timing ephemeris around the initial parameters, allowing us to search for small linear departures alongside noise processes and GW signals. For a single observation this can be written as

$$t_{\text{det},i}(\vec{\beta}) = t_{\text{det},i}(\vec{\beta}_0) + \left[\sum_j \left. \frac{\partial t_{\text{det},i}}{\partial \beta_j} \right|_{\vec{\beta}_0} \times (\beta_j - \beta_{0,j}) \right], \tag{7.3}$$

and for the full deterministic vector we write this compactly as

$$\vec{t}_{\text{det}}(\vec{\beta}) = \vec{t}_{\text{det}}(\vec{\beta}_0) + M\vec{\epsilon}, \tag{7.4}$$

where M is an $(N \times m)$ matrix of partial derivatives of the TOAs with respect to each timing-ephemeris parameter (evaluated at the initial fitting solution,

---

[1]For a discussion of modifying the PTA data model in the cases of non-Gaussianity or non-stationarity, see (2) and (3), respectively.

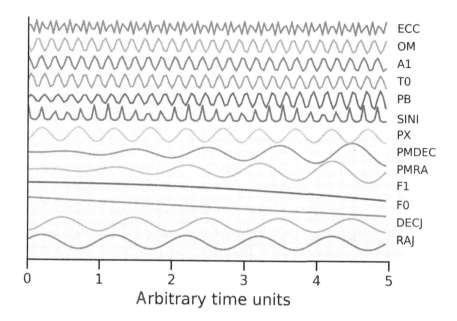

Figure 7.1: The columns of M in Eq. 7.5 are treated as a basis on which to expand timing-ephemeris perturbations in terms of different parameters. The time-dependent behavior of these basis vectors are shown here. Some timing-ephemeris offsets are clearly (quasi-)periodic in time, such as those due to positional and proper motion uncertainties. Adapted from https://github.com/nanograv/pulsar_timing_school.

and referred to as the *design matrix*), and $\vec{\epsilon}$ is a vector of linear parameter offsets from the initial fitting solution. For a simple timing ephemeris that involves a constant offset and quadratic spindown, the design matrix takes the form

$$M = \begin{pmatrix} 1 & t_1 & t_1^2 \\ 1 & t_2 & t_2^2 \\ \vdots & \vdots & \vdots \\ 1 & t_N & t_N^2 \end{pmatrix}. \tag{7.5}$$

An example of the time-dependent behavior of these quadratic spindown terms, as well as other astrometric and pulsar-binary terms, are shown in Fig. 7.1.

The design matrix typically contains columns with disparate dynamic ranges of the elements. This could lead to numerical instabilities when evaluating some of the quantities introduced later. We can easily resolve this by creating a stabilized version of M where each column is divided by its L^2 norm. Alternatively, we can perform an SVD (singular value decomposition) on M such that $M = USV^{\mathrm{T}}$. The matrix $U = [G_c\ G]$ can be partitioned into column spaces corresponding to the first m non-singular components, G_c (an $(N \times m)$ matrix), and the singular null-space components, G (an $(N \times (N - m))$ matrix). The design matrix can then be replaced with G_c, or the L^2-stabilized M, in all subsequent inference. However, care should be taken to store all quantities associated with stabilizing M in order to transform the inferred timing-ephemeris deviations back to their physical values (if relevant).

Prior: The timing ephemeris parameters are very well-constrained by pulsar-timing observations; their inference is likelihood dominated. Hence, we usually place an improper uniform prior on these small linear departures, $\vec{\epsilon}$, which is equivalent to a zero-mean Gaussian prior of infinite variance:

$$p(\vec{\epsilon}) = N(\text{Mean} = \vec{0}, \text{Variance} = \infty). \tag{7.6}$$

Introducing infinities seems like a dangerous prospect, but as we discuss later, we never actually have to worry about this in practice (1).

7.1.2 Achromatic Low-frequency Processes

Millisecond pulsars are very stable rotators, but they're not perfect. It is well known (e.g., 4) that pulsars can suffer from intrinsic instabilities that cause quasi-random-walk behaviour in pulsar pulse phase, period, or spindown rate. The result appears as long-timescale noise processes, often called *red noise* since the noise power is predominately at low sampling frequencies of the timing residuals[2]. Some example time-domain red noise realizations are shown in Fig. 7.2. We characterize this intrinsic red noise (and indeed any low-frequency process that has no radio-frequency dependence, including the GWB) in the timing residuals using a Fourier basis. Consider the Fourier series for an arbitrary function of time, $f(t)$:

$$f(t) = \frac{1}{2}a_0 + \sum_{k=1}^{\infty} a_k \sin(2\pi k t/T) + \sum_{k=1}^{\infty} b_k \cos(2\pi k t/T), \tag{7.7}$$

where a_0, a_k, b_k are Fourier coefficients, and k indexes the harmonics of the base sampling frequency $1/T$ that is the inverse of the data span.

In PTA analysis, we lose all sensitivity to constant offsets such as a_0 by fitting for constant phase offsets. We are also only interested in describing low-frequency behavior, so we can truncate the harmonic sum to some reasonable frequency, N_f. What remains of this expression can be written in more compact form for the vector of delays induced by low-frequency processes as

$$\vec{t}_{\mathrm{red}} = F\vec{a}, \tag{7.8}$$

---

[2]To understand why we call things "white noise", "red noise", etc., consider visible light. White light has equal power in all components of visible light, just as white noise has equal power across all sampling frequencies. By contrast, red light has an excess of lower frequency red components, as does red noise.

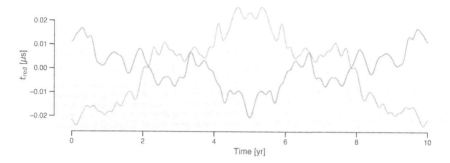

Figure 7.2: Example time-domain realizations of red noise, with $A = 10^{-15}$ and $\gamma = 2.5$. Adapted from https://github.com/nanograv/pulsar_timing_school.

where $\vec{a} = (a_1, b_1, a_2, b_2, \ldots, a_{N_f}, b_{N_f})^{\mathrm{T}}$ and F is the $(N \times 2N_f)$ *Fourier design matrix* that corresponds to columns of alternating sines and cosines for each frequency evaluated at the different observation times

$$
F = \begin{pmatrix}
\sin(2\pi t_1/T) & \cos(2\pi t_1/T) & \cdots & \sin(2\pi N_f t_1/T) & \cos(2\pi N_f t_1/T) \\
\sin(2\pi t_2/T) & \cos(2\pi t_2/T) & \cdots & \sin(2\pi N_f t_2/T) & \cos(2\pi N_f t_2/T) \\
\vdots & \vdots & \vdots & \vdots & \vdots \\
\sin(2\pi t_N/T) & \cos(2\pi t_N/T) & \cdots & \sin(2\pi N_f t_N/T) & \cos(2\pi N_f t_N/T)
\end{pmatrix}.
$$
(7.9)

Lower frequency red-noise effects are highly covariant with the quadratic spin-down of the timing ephemeris, so modeling below $1/T$ is not necessary. We call this limitation to low-frequency sensitivity the "quadratic wall". However, some techniques have been developed to improve the modeling of very red processes by including frequencies below $1/T$ (5).

Prior: We model these achromatic red noise terms as zero-mean Gaussian random processes

$$
p(\vec{a}|\vec{\eta}) = \frac{\exp\left(-\frac{1}{2}\vec{a}^{\mathrm{T}}\phi^{-1}\vec{a}\right)}{\sqrt{\det(2\pi\phi)}},
$$
(7.10)

where $\langle \vec{a}\,\vec{a}^{\mathrm{T}} \rangle = \phi$, and $\vec{\eta}$ are hyper-parameters of the Gaussian variance on these processes. The covariance matrix of the Fourier coefficients will include all potential achromatic low-frequency processes, which necessarily also includes the GWB and any other spatially-correlated noise processes. Hence,

$$
[\phi]_{(ak)(bj)} = \Gamma_{ab}\rho_k\delta_{kj} + \kappa_{ak}\delta_{kj}\delta_{ab},
$$
(7.11)

where (a, b) index over pulsars, (k, j) index over sampling frequencies of the timing residuals, and Γ_{ab} is the GWB overlap reduction function coefficient for pulsars (a, b) (or the equivalent cross-correlation coefficient for spatially-correlated noise processes). The terms ρ_k / κ_{ak} are related to the power spectral density (PSD) of the timing delay, $S(f)$ (units of time^3), induced by the GWB (or spatially-correlated noise process) and intrinsic per-pulsar red noise, respectively, such that $\rho(f) = S(f)\Delta f$, and $\Delta f = 1/T$ (and likewise for $\kappa_a(f)$).

There is alot of flexibility in how we can model the PSD of these processes. The default assumption is for a power-law GWB and/or intrinsic per-pulsar red noise, such that

$$\rho(f) = \frac{h_c(f)^2}{12\pi^2 f^3}\frac{1}{T} = \frac{A^2}{12\pi^2}\frac{1}{T}\left(\frac{f}{1\,\mathrm{yr}^{-1}}\right)^{-\gamma}\mathrm{yr}^2, \qquad (7.12)$$

where $h_c(f) = A(f/1\,\mathrm{yr}^{-1})^\alpha$ is the GWB characteristic strain spectrum, $\gamma \equiv 3 - 2\alpha$, and

$$\kappa_a(f) = \frac{A_a^2}{12\pi^2}\frac{1}{T}\left(\frac{f}{1\,\mathrm{yr}^{-1}}\right)^{-\gamma_a}\mathrm{yr}^2. \qquad (7.13)$$

In these parametrizations, the power-law amplitude, A, is referenced to a frequency of an inverse year. There are many other parametrized spectrum representations that are in common usage, including: a turnover spectrum to encapsulate the influence of non-GW final-parsec dynamical influences on SMBBH evolution (6); spectra with high-frequency knees to account for SMMBH population finiteness (7); astrophysics-driven spectral parametrizations (8; 9; 10; 11); t-process spectra that allow for some degree of fuzziness around a power-law (12); and finally the free-spectrum model that is agnostic to the spectral shape, allowing complete per-frequency flexibility in the modeling (13). In all of these we care more about recovering the hyper-parameters, $\vec{\eta}$ (e.g., A, γ), than the actual Fourier coefficients of the process.

7.1.3 Chromatic Low-frequency Processes

Pulsars emit radio pulses that propagate through interstellar space on the way to our observatories on Earth. This space is filled with ionized material, leading to radio-frequency dependent (i.e., chromatic) propagation delays for different components of the pulse. Components at lower radio frequencies will experience greater delays than those at higher frequencies. The leading-order chromatic effect is *dispersion*, where the group delay of a given radio-frequency component, ν_{obs}, is $t_{\mathrm{delay}} = (\mathrm{DM}/K)(1/\nu_{\mathrm{obs}}^2)$, where the dispersion constant is defined to be $K = 2.41 \times 10^{-16}\,\mathrm{Hz}^{-2}\mathrm{cm}^{-3}\mathrm{pc}\,\mathrm{s}^{-1}$, and the *dispersion measure*, DM, is defined to be the line-of-sight integrated column density of free electrons

$$\mathrm{DM} = \int_0^L n_e\,dl. \qquad (7.14)$$

Leading-order dispersive effects are modeled in the timing ephemeris, however, variations in the electron column density can induce chromatic red noise in the timing residuals. Example time-domain realizations of chromatic delays induced by DM variations are shown in Fig. 7.3. There are variety of techniques

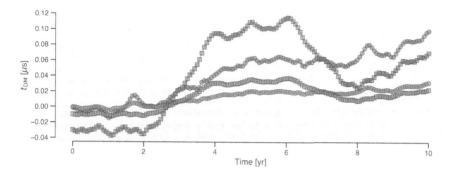

Figure 7.3: Example time-domain realizations of chromatic delays induced by dispersion-measure variations, with $A = 10^{-15}$ and $\gamma = 2.5$. There are two realizations of delays for two different radio-frequency components (indicated by the line colors). Adapted from https://github.com/nanograv/pulsar_timing_school.

used to model these DM variations (e.g., 14; 15), but we focus the discussion here on the two most prominent.

DMX: This model is used by the NANOGrav Collaboration, where DM variations are treated as an epoch-by-epoch offset in DM. The DM offset is fit alongside all other timing-ephemeris parameters as a purely deterministic effect.

DM GP: This model treats DM variations as a chromatic red-noise (Gaussian) process. This is written as

$$\vec{t}_{DM} = F_{DM}\vec{a}_{DM}, \tag{7.15}$$

where each element of F_{DM} is related to that of F via

$$F_{DM}(f_k, t_i) = F(f_k, t_i) \times 1/(K\nu_{obs,i}^2), \tag{7.16}$$

and $\nu_{obs,i}$ is the observed radio frequency of the ith TOA. Unlike achromatic red noise, this DM GP has no quadratic function of the time-series to act as a proxy for low-frequency power. Hence two additional parameters (DM1 and DM2) are added to the deterministic timing ephemeris, corresponding to coefficients of chromatic linear and quadratic trends in the TOAs. Note also that, while we have assumed a Fourier basis for the DM variations here, one could also consider a coarse-grained time-domain basis to allow other kinds of GP kernels to model quasi-periodic features and even radio-frequency-band evolution of DM variations[3].

---

[3]The Kernel Cookbook: https://www.cs.toronto.edu/ duvenaud/cookbook

Dispersion may be the leading-order chromatic effect on the TOAs, but is by no means the only one. There are a litany of other chromatic influences on pulse arrival times as they propagate through the ionized interstellar medium (16), such as turbulence-induced pulse broadening, which can lead to $\sim 1/\nu_{\text{obs}}^{4.4}$ dependent delays. These influences are easily modeled through additional terms like Eq. 7.15.

Prior: Under the DMX model, DM variations are part of the timing ephemeris, and treated accordingly in the prior assumption. When modeling DM variations as a Gaussian process, the prior treatment is very similar to achromatic processes, where $\langle \vec{a}_{\text{DM}}\, \vec{a}_{\text{DM}}^{\text{T}} \rangle = \phi_{\text{DM}}$, and

$$[\phi_{\text{DM}}]_{(ak)(bj)} = \lambda_{ak}\delta_{kj}\delta_{ab}, \tag{7.17}$$

where $\lambda(f) = S_{\text{DM}}(f)\Delta f$. The PSD of DM variations can be modeled with all of the flexibility of achromatic processes, but typically a power-law is adopted with a minor modification in the definition:

$$\lambda_a(f) = \frac{A_{\text{DM,a}}^2}{T}\left(\frac{f}{1\,\text{yr}^{-1}}\right)^{-\gamma_{\text{DM},a}}\,\text{yr}^2. \tag{7.18}$$

7.1.4 White Noise

White noise has a flat power spectral density across all sampling frequencies. It therefore does not exhibit long-timescale trends in the residual time-series. An example time-domain covariance matrix from the different white noise sources mentioned below is shown in Fig. 7.4. White noise also exhibits no inter-pulsar correlations. There are three main sources of white noise in PTA analysis:

Radiometer noise & EFAC: All pulses observed within a given data-taking epoch ($\sim$20 – 30 minutes) are de-dispersed and folded over the pulsar's spin period, which is then fit to a long-timescale averaged pulse template to compute the actual TOA. The standard TOA uncertainties from radiometer noise are merely template-fitting uncertainties. But not all sources of fitting uncertainty may be propagated into the final quoted TOA uncertainty, leading us to introduce a correction factor, or *Extra FACtor (EFAC)* that acts as a multiplicative correction to the uncertainties. We apply these EFACs to unique combinations of telescope receivers and backends (a given unique combination is denoted a "system" (17))

$$\langle n_{i,\mu}n_{j,\nu} \rangle = F_\mu^2 \sigma_i^2 \delta_{ij}\delta_{\mu\nu}, \tag{7.19}$$

where $n_{i,\mu}$ is the timing delay induced by white noise at observation i in receiver-backend system μ; σ_i is the TOA uncertainty for observation i; and F_μ is the EFAC for μ. EFACs that are significantly different from unity would raise suspicions of data quality.

EQUAD: There may be additional white noise that is completely separate from radiometer noise, from e.g., instrumental effects. To account for additional non-multiplicative corrections to the TOA uncertainties, we include an *Extra QUADrature (EQUAD)* noise term. This EQUAD adds in quadrature to the term above, which now becomes

$$\langle n_{i,\mu} n_{j,\nu} \rangle = F_\mu^2 \sigma_i^2 \delta_{ij} \delta_{\mu\nu} + Q_\mu^2 \delta_{ij} \delta_{\mu\nu}, \tag{7.20}$$

and also applies to unique instrument systems.

Pulse phase jitter & ECORR: Within a given observing *epoch*, a finite number of pulses are being folded and fit to a standard pulse profile template. This leads to an effect known as *pulse phase jitter* that contributes an additional white noise term; essentially the folding of a finite number of pulses from epoch to epoch leaves some residual shape fluctuation with respect to the profile template. While this depends on the number of pulses recorded in a given observing epoch, this is typically not changing much over the span of the pulsar's dataset, and so it is again just discriminated by unique receiver-backend systems. What's more, in a given observing epoch NANOGrav typically records many near-simultaneous TOAs across neighboring radio-frequency bands. Given that these correspond to the same folded pulses, the pulse phase jitter will be fully correlated across these bands. This gives us *Extra Correlated (ECORR)* white noise, which is uncorrelated between different observing epochs, but fully correlated between different bands within the same epoch.

ECORR can be treated in two different ways within our pipelines. An important piece of initial book-keeping involves making sure that our TOAs are sorted by epoch. We can then either,

(a) Model ECORR using a low-rank basis (similar to the Fourier basis for low-frequency processes, but now on a basis of epochs), such that

$$\vec{n}_{\text{ECORR}} = U \vec{j}, \tag{7.21}$$

where $\vec{j}$ is a jitter vector of length N_{epoch}, and U is an $(N \times N_{\text{epoch}})$ matrix that has entry values of 1 when a TOA row lies within an epoch column, and zero everywhere else, e.g.,

$$
U = \begin{pmatrix}
1 & & & & \\
\vdots & & & & \\
1 & & & & \\
& 1 & & & \\
& \vdots & & & \\
& 1 & & & \\
& & \ddots & & \\
& & & 1 & \\
& & & \vdots & \\
& & & 1 &
\end{pmatrix}. \tag{7.22}
$$

$\underbrace{}_{N_{\text{epoch}}}$

U is sometimes referred to as the *epoch exploder matrix*, given that it explodes epoch assignments out to the full rank of TOAs.

Prior: We treat jitter/ECORR in this sense as a Gaussian process that is, fully correlated within an epoch but uncorrelated across different epochs. As such our prior is

$$
p(\vec{j}|\mathcal{J}) = \frac{\exp\left(-\frac{1}{2}\vec{j}^{\mathrm{T}}\mathcal{J}^{-1}\vec{j}\right)}{\sqrt{\det(2\pi\mathcal{J})}}, \tag{7.23}
$$

where $\langle\vec{j}\,\vec{j}^{\mathrm{T}}\rangle = \mathcal{J}$, and $\vec{J}$ is a vector of jitter rms values for distinct receiver-backend systems that parametrize the jitter covariance matrix $\mathcal{J}$.

(b) Model ECORR alongside other white noise processes like radiometer noise and EQUAD. This is a tempting approach, because we are not actually interested in what $\vec{j}$ is, in the same way that we are not interested in what the time-series of other white-noise timing delays are. Jitter/ECORR can be packaged alongside EFAC and EQUAD as simply another white noise term

$$
\langle n^J_{i,\mu} n^J_{j,\nu}\rangle = J^2_\mu \delta_{e(i)e(j)}\delta_{\mu\nu}, \tag{7.24}
$$

where $e(i)$ indexes the epochs of each observation such that $\delta_{e(i)e(j)}$ is zero unless TOAs lie within the same epoch, and as before μ indexes each receiver-backend system.

7.2 THE LIKELIHOOD

We now have all the main ingredients with which to piece together the PTA likelihood. To keep things relatively straightforward when introducing this

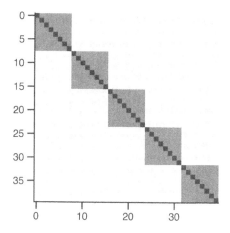

Figure 7.4: Example time-domain covariance matrix of white noise. Noise from EFAC-scaled TOA uncertainties and EQUAD contributions are uncorrelated with other observations, manifesting as a diagonal covariance matrix (dark blue). Jitter/ECORR is uncorrelated between different epochs, but fully correlated within epochs, giving a block-diagonal covariance matrix (light blue). Adapted from https://github.com/nanograv/pulsar_timing_school.

likelihood, we model the residuals as being only the sum of contributions from linear timing ephemeris deviations (containing the DMX model of DM variations), achromatic low-frequency processes (including a GWB), and various white noise processes[4]:

$$\vec{\delta t} = \boldsymbol{M}\vec{\epsilon} + \boldsymbol{F}\vec{a} + \vec{n}, \tag{7.25}$$

where $\vec{n}$ contains jitter noise. We form a new vector of noise- and signal-mitigated timing residuals that acts as a model-dependent estimate of $\vec{n}$, such that

$$\vec{r} = \vec{\delta t} - \boldsymbol{M}\vec{\epsilon} - \boldsymbol{F}\vec{a}. \tag{7.26}$$

Since $\vec{r}$ acts as our estimate of Gaussian white noise, its likelihood is simple:

$$p(\vec{r}|\vec{\epsilon}, \vec{a}, \dots, \vec{\eta}) = \frac{\exp\left(-\frac{1}{2}\vec{r}^{\mathrm{T}}\boldsymbol{N}^{-1}\vec{r}\right)}{\sqrt{\det(2\pi\boldsymbol{N})}}, \tag{7.27}$$

where $\vec{\eta}$ are hyper-parameters that (for now) just appear as EFACs, EQUADs, ECORRs etc., in the white noise covariance matrix, $\boldsymbol{N}$.

For convenience, we now group all rank-reduced processes (e.g., linear timing-ephemeris deviations, low-frequency processes) together, such that

$$\vec{r} = \vec{\delta t} - \boldsymbol{T}\vec{b}, \tag{7.28}$$

---

[4]We will consider other deterministic processes later, such as chromatic features in individual pulsars, and resolvable individual SMBBH signals across the entire PTA.

where $T = [M\ F]$, and $\vec{b} = \begin{bmatrix} \vec{\epsilon} \\ \vec{a} \end{bmatrix}$. This kind of compact representation is easily expanded with additional model compoenents, whether they be the Gaussian process model of DM variations ($F_{DM}\vec{a}_{DM}$), or the epoch-expanded model of jitter/ECORR ($U\vec{j}$). All you need to do is concatenate the relevant basis matrices onto T, and likewise the coefficient vectors onto $\vec{b}$.[5]

What now? Well, we have a likelihood, and in principle we could proceed with inference on our pulsar-timing residuals using just Eq. 7.27. That would be a grave mistake though; we would likely learn very little about all these various coefficient vectors in $\vec{b}$ or the white noise parameters. That's because we have forgotten to include a proper prior on $\vec{b}$. At the moment the prior is implicitly uniform over the unbounded range $[-\infty, +\infty]$. This is appropriate for the timing-ephemeris deviations, since the inference on those coefficients will be data-dominated. But all other processes are exceptionally small, requiring proper control with well-chosen priors. Looking back to the various descriptions of each process contributing to the pulse time-series, the answer is straightforward: we assume these are are zero-mean random Gaussian processes with parametrized variances. The prior on $\vec{b}$ is then

$$p(\vec{b}|\vec{\eta}) = \frac{\exp\left(-\frac{1}{2}\vec{b}^{T}B^{-1}\vec{b}\right)}{\sqrt{\det(2\pi B)}}, \tag{7.29}$$

where

$$B = \begin{pmatrix} \infty & 0 \\ 0 & \phi \end{pmatrix}, \tag{7.30}$$

and $\vec{\eta}$ are parameters that control the behaviour of $B \equiv B(\vec{\eta})$. However, given that these are parameters of a prior, we call these *hyper-parameters*. Once again, this prior matrix B is easily expanded to include more blocks like ϕ_{DM} or J. We'll see next how we can piece together the likelihood and this prior to get the full hierarchical PTA likelihood. Finally, don't worry about that infinity block! As we'll see soon, we only ever have to deal with the inverse of B, making this infinity drop to zero.

7.2.1 Full hierarchical Likelihood

The joint probability density of the process coefficients and their variance hyper-parameters can be written as a chain of conditional probabilities

$$p(\vec{b}, \vec{\eta}|\vec{\delta t}) \propto p(\vec{\delta t}|\vec{b}) \times p(\vec{b}|\vec{\eta}) \times p(\vec{\eta}), \tag{7.31}$$

---

[5]There is a notable alternative to the T-matrix method, called the G-matrix method (18), where G is the null space of the timing ephemeris design matrix, M (see Sec. 7.1.1). This method is appropriate for when you don't care about the timing ephemeris at all, and simply want to marginalize over it. When constructing a statistic or writing a likelihood, you just replace all vectors $\vec{x}$ by $G^{T}\vec{x}$, and all matrices X by $G^{T}XG$. In so doing, you are projecting these quantities into the null space of the timing ephemeris, which can be shown to be mathematically equivalent to marginalization (1) (see also https://gwic.ligo.org/assets/docs/theses/taylor_thesis.pdf).

where $p(\vec{\eta})$ is the prior density on the hyper-parameters. The first two terms in this chain give the hierarchical PTA likelihood, $p(\vec{\delta t}|\vec{b}, \vec{\eta}) = p(\vec{\delta t}|\vec{b}) \times p(\vec{b}|\vec{\eta})$. Let's see what this looks like when written out in full:

$$p(\vec{\delta t}|\vec{b}, \vec{\eta}) = \frac{\exp\left(-\frac{1}{2}(\vec{\delta t} - \boldsymbol{T}\vec{b})^{\mathrm{T}} \boldsymbol{N}^{-1}(\vec{\delta t} - \boldsymbol{T}\vec{b})\right)}{\sqrt{\det(2\pi\boldsymbol{N})}} \times \frac{\exp\left(-\frac{1}{2}\vec{b}^{\mathrm{T}}\boldsymbol{B}^{-1}\vec{b}\right)}{\sqrt{\det(2\pi\boldsymbol{B})}}. \quad (7.32)$$

For a single pulsar, this doesn't look too scary, and in fact it can be pretty quick to evaluate on modern laptops. The speed with which we can evaluate this actually depends on how jitter has been treated. If jitter is packaged along with other rank-reduced processes in $\boldsymbol{T}\vec{b}$, then $\boldsymbol{N}$ is a diagonal matrix that is trivial to invert and whose determinant is simply the product of diagonal elements. Likewise, for a single pulsar $\boldsymbol{B}$ is a relatively small and easily-invertable diagonal matrix, such that $\boldsymbol{B}^{-1} = \mathrm{diag}(\boldsymbol{0}, \phi^{-1})$ and the diagonal matrix ϕ is trivial to invert. Using the fact that the inverse of a matrix determinant is equivalent to the determinant of the matrix's inverse, we see that the infinity block poses no practical problems.

What about when jitter is instead modeled inside $\boldsymbol{N}$? This white noise covariance matrix then becomes *block diagonal*, with a block for each epoch. The inverse is simply another block diagonal matrix formed of each inverted block. Let's consider each block to be of the form $\boldsymbol{N}_e = \boldsymbol{D}_e + J_e^2 \vec{u}_e \vec{u}_e^{\mathrm{T}}$, where $\boldsymbol{D}_e$ is the usual diagonal part containing radiometer noise, EFAC, and EQUAD contributions; J_e is the jitter rms in this epoch; and $u_e = (1, 1, \ldots, 1)^{\mathrm{T}}$ contains as many entries as there are TOAs in this epoch. The inverse of this block is efficiently computed using the *Sherman-Morrison formula* (19):

$$\boldsymbol{N}_e^{-1} = \boldsymbol{D}_e^{-1} - \frac{\boldsymbol{D}_e^{-1}\vec{u}_e\vec{u}_e^{\mathrm{T}}\boldsymbol{D}_e^{-1}}{J_e^{-2} + \vec{u}_e^{\mathrm{T}}\boldsymbol{N}_e^{-1}\vec{u}_e}. \quad (7.33)$$

Note that the denominator of the right-most term is a scalar quantity. The Sherman-Morrison formula is a special case of the *Woodbury matrix identity* that we will see more of soon.

For a full PTA, Eq. 7.32 is modified to become a product over the $p(\vec{\delta t}|\vec{b})$ terms, and controlled by a prior on $\vec{b}$ that allows for inter-pulsar correlations induced by a GWB or systematic noise processes:

$$p(\{\vec{\delta t}\}|\{\vec{b}\}, \vec{\eta}) = \left[\prod_{a=1}^{N_p} p(\vec{\delta t}_a|\vec{b}_a)\right] \times p(\{\vec{b}\}|\vec{\eta}), \quad (7.34)$$

where $p(\{\vec{b}\}|\vec{\eta})$ has the same form as Eq. 7.29 with $\vec{b}$ replaced by the concatenation of coefficient vectors over all pulsars, and with the $\boldsymbol{B}$ matrix becoming a matrix of $N_p \times N_p$ blocks for each pair of pulsars. For cross-pairings of pulsars, the usual infinity block becomes zero since the pulsar timing ephemerides are uncorrelated, and the ϕ block contains only contributions from inter-pulsar correlated processes (see Eq. 7.11). $\boldsymbol{B}$ is now a large band diagonal matrix, which still allows for straightforward and efficient computation of its inverse and determinant. We do so by temporarily reordering the matrix into an $N_b \times N_b$ matrix of blocks for each $\vec{b}$ coefficient; this makes $\boldsymbol{B}$ a block diagonal matrix that we can easily invert one block at a time, where upon inverting each block we slot the elements back into their original band diagonal structure to return $\boldsymbol{B}^{-1}$.

Equation 7.34 has a *very* large parameter space to explore! For each pulsar there are timing ephemeris deviations, intrinsic red noise coefficients and hyper-parameters, EFACs, EQUADs, and ECORRs, plus potential chromatic coefficients plus hyper-parameters, jitter vectors, and any other parameters that are used to model features within the residual time series. Add to that the coefficients and hyper-parameters of inter-pulsar correlated processes like the GWB, and we have a parameter space that encroaches on ~1000 dimensions or more for currently-sized PTAs of ~50 pulsars. Unguided Metropolis-Hastings-based MCMC is going to have big difficulties in sampling from this parameter space, for reasons that include the necessary burn-in time, auto-correlation lengths, and the well-known *Neal's funnel* conundrum of trying to sample coefficient parameters from distributions whose variances are also being updated (20). However, recent advances in Gibbs sampling (where parameter blocks are updated through sequential conditional distribution draws) (1), and Hamiltonian MCMC (where likelihood gradient information is used to guide the chain exploration) (13; 21) have made sampling from this hierarchical likelihood tractable.

7.2.2 Marginalized Likelihood

Let's ask ourselves what we really care about most of the time in our noise or GW analyses. Is it the particular coefficients that describe the one realization of noise or GW signal that we see in the residuals, or is it the statistical properties of these processes? We're dealing with stochastic processes, so the vast majority of the time it is the latter. Hence, we need not trouble ourselves with numerically sampling from this huge parameter space of $\vec{b}$ coefficients plus hyper-parameters, when all we care about are the hyper-parameters. We can use the wonderful integration properties of chained Gaussian distributions to *analytically* marginalize over these $\vec{b}$ coefficients, leaving a marginalized likelihood that depends only on the hyper-parameters (13; 18). This has the form

$$p(\{\vec{\delta t}\}|\vec{\eta}) = \int p(\{\vec{\delta t}\}|\{\vec{b}\}, \vec{\eta})\, d^{N_p}\vec{b} = \frac{\exp\left(-\frac{1}{2}\vec{\delta t}^{\mathrm{T}} \boldsymbol{C}^{-1} \vec{\delta t}\right)}{\sqrt{\det(2\pi\boldsymbol{C})}}, \qquad (7.35)$$

where $\boldsymbol{C} = \boldsymbol{N} + \boldsymbol{T}\boldsymbol{B}\boldsymbol{T}^{\mathrm{T}}$. Now, to get to this compact form, we have implicitly used the very powerful *Woodbury matrix identity* (22). This simplifies the inversion of the large dense matrix $\boldsymbol{C}$ such that

$$\boldsymbol{C}^{-1} = (\boldsymbol{N} + \boldsymbol{T}\boldsymbol{B}\boldsymbol{T}^{\mathrm{T}})^{-1} = \boldsymbol{N}^{-1} - \boldsymbol{N}^{-1}\boldsymbol{T}\boldsymbol{\Sigma}^{-1}\boldsymbol{T}^{\mathrm{T}}\boldsymbol{N}^{-1}, \qquad (7.36)$$

where $\mathbf{\Sigma} = \mathbf{B}^{-1} + \mathbf{T}^\mathrm{T}\mathbf{N}^{-1}\mathbf{T}$. This may not look simpler, but it is! It's also much faster to compute. Performing a Cholesky decomposition (23) to invert a symmetric positive-definite matrix like $\mathbf{C}$ usually scales as $\mathcal{O}(N_p^3 N_\mathrm{TOA}^3)$.[6] But in Eq. 7.36, the bottleneck operation is $\mathbf{\Sigma}^{-1}$, which still requires a Cholesky inversion, but which now scales as $\mathcal{O}(N_p^3 N_b^3)$. The number of $\vec{b}$ coefficients for each pulsar is typically much less than the number of TOAs, rendering this a significant acceleration. The determinant of $\mathbf{C}$ is also made more tractable by the Woodbury identity, such that $\det(\mathbf{C}) = \det(\mathbf{N})\det(\mathbf{B})\det(\mathbf{\Sigma})$.

It's important to note what the form of $\mathbf{TBT}^\mathrm{T}$ actually means for the rank-reduced processes. If we simply look at the achromatic process sector of this dense matrix to inspect the covariance between times t and $(t + \tau)$ in pulsars a and b, we get

$$[\mathbf{TBT}^\mathrm{T}]_{(ab),\tau} = \sum_k^{N_f} [\boldsymbol{\phi}]_{ab} \cos(2\pi k\tau/T), \qquad (7.37)$$

where $[\boldsymbol{\phi}]_{ab} = \Gamma_{ab}\rho/T$, Γ_{ab} is the cross-correlation coefficient between pulsars, ρ is the PSD of the process, and T is the observational timing baseline. In this form we see that Eq. 7.37 is simply the discretized form of the *Wiener-Khinchin theorem* (24; 25) to translate the PSD of a stochastic process into its temporal covariance.

Equation 7.36 is how we practically implement the form of Eq. 7.35 within production-level GW search pipelines, the most prominent of which is ENTERPRISE (Enhanced Numerical Toolbox Enabling a Robust PulsaR Inference SuitE)[7]. We use this marginalized likelihood with all of the numerical Bayesian techniques outlined in Chapter 6, including parameter estimation of GW signals and noise processes, upper limits on signal parameters[8], and model selection. Given that we expect the stochastic GWB will be the first manifestation of GW detection in our PTAs, the model selection of primary current interest is between a spatially-correlated GWB signal and that of a spatially-uncorrelated common-spectrum red process. The only difference between these models is the presence of GWB-induced inter-pulsar correlations as described by the Hellings & Downs function, since these cross-correlations are the undeniable fingerprint of the influence of GWs on our pulsar-timing experiment.

A graphical representation of a PTA search for the SGWB is shown in Fig. 7.5 as a Bayesian network. This network illustrates the chain of conditional statistical dependencies of all processes that constitute our model of the data.

---

[6]A Cholesky decomposition factors a symmetric positive-definite matrix into $\mathbf{X} = \mathbf{LL}^\mathrm{T}$ such that $\mathbf{L}$ is a lower triangular matrix, making equation solving and inversion simpler to evaluate. Furthermore, the determinant of $\mathbf{X}$ is simply the product of the squares of the diagonal elements of $\mathbf{L}$.

[7]enterprise: https://github.com/nanograv/enterprise

[8]In Bayesian inference the upper limit is just the parameter value at the required percentile of the 1D-marginalized parameter posterior distribution.

7.2.3 Modeling Deterministic Signals

Thus far, we have talked about modeling stochastic processes, either signal or noise. However, we also want to be able to model signals for which we have a deterministic description of the time-domain behavior, e.g., some of the sources mentioned in Chapter 4 – individually resolvable SMBBH signals; GW bursts; Solar System ephemeris perturbations; or even a full non-linear pulsar timing ephemeris analysis. Fortunately, it is incredibly easy to slot this kind of signal description into the PTA likelihood framework. We recall that the likelihood is meant to be a probabilistic description of the random Gaussian components of the residual time series. Way back in Eq. 7.26, we created a time-series of our best estimate of the Gaussian white noise in our data. To model additional deterministic processes, we simply replace $\vec{r}$ in Eq. 7.26 with $\vec{r} \rightarrow \vec{r} - \vec{s}(\vec{\theta})$, where $\vec{s}(\theta)$ is a deterministic signal function that depends on parameters $\vec{\theta}$. For the marginalized likelihood of Eq. 7.35, this just means we make the following replacement

$$\vec{\delta t} \rightarrow \vec{\delta t} - \vec{s}(\vec{\theta}). \tag{7.38}$$

7.3 LIKELIHOOD-BASED STATISTICS

The full form of the PTA likelihood is a sufficient statistic for all kinds of deterministic signal and stochastic process searches. However there are many statistics that can be derived from this PTA likelihood for the purpose of more limited searches or frequentist constraints. The following are some of the most used.

7.3.1 GWB Statistics

7.3.1.1 Optimal Statistic

The optimal statistic (OS) is a frequentist estimator of the amplitude of an isotropic stochastic GWB (26; 27; 28). It can be derived by maximizing the PTA likelihood under a first-order expansion around the Hellings & Downs inter-pulsar correlation coefficients (29). Given that $\boldsymbol{C} = \langle \vec{\delta t} \vec{\delta t}^{\mathrm{T}} \rangle$ is the time-domain covariance of residuals between arbitrary pairs of pulsars, let us introduce the following labels for the following specific combinations:

$$\begin{aligned} \boldsymbol{P}_a &= \boldsymbol{C}_{aa}, \\ \boldsymbol{S}_{ab} &= A_{\mathrm{GWB}}^2 \tilde{\boldsymbol{S}}_{ab} = \boldsymbol{C}_{ab}. \end{aligned} \tag{7.39}$$

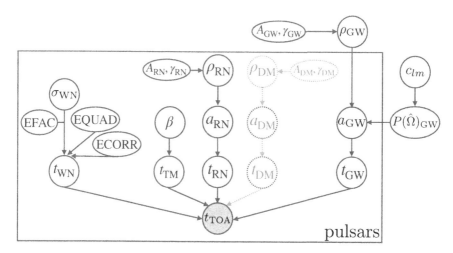

Figure 7.5: Probabilistic graphical model (or Bayesian network) for a PTA search for a SGWB. Arrows indicate the chain of conditional statistical dependencies. All processes inside the box are intrinsic to each pulsar, implying that the joint probability distribution of data is given by the product of probabilities over all pulsars. Arrows outside the box indicate priors on inter-pulsar correlated processes, which in this case is the frequency and angular spectral behaviour of the SGWB. Dashed arrows and circles modify this model to treat DM variations as a random Gaussian process, rather than part of the timing ephemeris. Additional deterministic per-pulsar (transient chromatic features) or common signals (Solar System ephemeris perturbations and individual GW signals) can be trivially added to this framework.

The first-order–expanded and marginalized PTA likelihood function can be written as (see Ref. (29) for full details)

$$\ln p(\{\vec{\delta t}\}|\vec{\eta}) \approx -\frac{1}{2}\left[\sum_{a=1}^{N_p}(\text{Tr}\ln \boldsymbol{P}_a + \vec{\delta t}_a^{\text{T}}\boldsymbol{P}_a^{-1}\vec{\delta t}_a) - \sum_{a=1}^{N_p}\sum_{b>a}^{N_p}\vec{\delta t}_a^{\text{T}}\boldsymbol{P}_a^{-1}\boldsymbol{S}_{ab}\boldsymbol{P}_b^{-1}\vec{\delta t}_b\right].$$
(7.40)

If we have performed a program of noise estimation for all pulsars, then the autocovariance matrices can all be assumed known, allowing us to form the log-likelihood ratio: $\ln \Lambda = \ln p(\{\vec{\delta t}\}|\vec{\eta}, \mathcal{H}_{\text{gw}}) - \ln p(\{\vec{\delta t}\}|\vec{\eta}, \mathcal{H}_{\text{noise}})$.[9] With all pulsar noise parameters known, we further assume the GWB characteristic strain spectrum takes its default power-law form with the exponent for a SMBBH population, $\alpha = -2/3$. This leaves only one parameter to be estimated: the amplitude of the characteristic strain spectrum at a frequency of $1/\text{yr}$, A_{GWB}. Evaluating this log-likelihood ratio gives

$$\ln \Lambda = \frac{A_{\text{GWB}}^2}{2}\sum_{a=1}^{N_p}\sum_{b>a}^{N_p}\vec{\delta t}_a^{\text{T}}\boldsymbol{P}_a^{-1}\tilde{\boldsymbol{S}}_{ab}\boldsymbol{P}_b^{-1}\vec{\delta t}_b.$$
(7.41)

It can also be shown that

$$\left\langle\sum_{a=1}^{N_p}\sum_{b>a}^{N_p}\vec{\delta t}_a^{\text{T}}\boldsymbol{P}_a^{-1}\tilde{\boldsymbol{S}}_{ab}\boldsymbol{P}_b^{-1}\vec{\delta t}_b\right\rangle = A_{\text{GWB}}^2\sum_{a=1}^{N_p}\sum_{b>a}^{N_p}\text{Tr}\left[\boldsymbol{P}_a^{-1}\tilde{\boldsymbol{S}}_{ab}\boldsymbol{P}_b^{-1}\tilde{\boldsymbol{S}}_{ba}\right].$$
(7.42)

Therefore, we can write the optimal estimator of the GWB amplitude as

$$\hat{A}^2 = \frac{\sum_{a=1}^{N_p}\sum_{b>a}^{N_p}\vec{\delta t}_a^{\text{T}}\boldsymbol{P}_a^{-1}\tilde{\boldsymbol{S}}_{ab}\boldsymbol{P}_b^{-1}\vec{\delta t}_b}{\sum_{a=1}^{N_p}\sum_{b>a}^{N_p}\text{Tr}\left[\boldsymbol{P}_a^{-1}\tilde{\boldsymbol{S}}_{ab}\boldsymbol{P}_b^{-1}\tilde{\boldsymbol{S}}_{ba}\right]},$$
(7.43)

thereby ensuring $\langle\hat{A}^2\rangle = A_{\text{GWB}}^2$. In the weak-signal or noise-only regime, $\langle\hat{A}^2\rangle = 0$ and the standard deviation of this estimator is

$$\sigma_0 = \left(\sum_{a=1}^{N_p}\sum_{b>a}^{N_p}\text{Tr}\left[\boldsymbol{P}_a^{-1}\tilde{\boldsymbol{S}}_{ab}\boldsymbol{P}_b^{-1}\tilde{\boldsymbol{S}}_{ba}\right]\right)^{-1/2}.$$
(7.44)

Using this, we can define a signal-to-noise ratio (SNR), corresponding to the number of standard deviations away from zero that the measured statistic is found to be:

$$\hat{\rho} = \frac{\hat{A}^2}{\sigma_0} = \frac{\sum_{a=1}^{N_p}\sum_{b>a}^{N_p}\vec{\delta t}_a^{\text{T}}\boldsymbol{P}_a^{-1}\tilde{\boldsymbol{S}}_{ab}\boldsymbol{P}_b^{-1}\vec{\delta t}_b}{\left(\sum_{a=1}^{N_p}\sum_{b>a}^{N_p}\text{Tr}\left[\boldsymbol{P}_a^{-1}\tilde{\boldsymbol{S}}_{ab}\boldsymbol{P}_b^{-1}\tilde{\boldsymbol{S}}_{ba}\right]\right)^{1/2}},$$
(7.45)

---

[9]The models being compared here are a GWB versus a spatially-uncorrelated common process, giving a statistic that acts as a frequentist proxy for the Bayesian odds of GWB-induced cross-correlations.

which has expectation value

$$\langle \rho \rangle = A_{\text{GWB}}^2 \left(\sum_{a=1}^{N_p} \sum_{b>a}^{N_p} \text{Tr} \left[\boldsymbol{P}_a^{-1} \tilde{\boldsymbol{S}}_{ab} \boldsymbol{P}_b^{-1} \tilde{\boldsymbol{S}}_{ba} \right] \right)^{1/2}. \tag{7.46}$$

Hence

$$\langle \ln \Lambda \rangle = \langle \rho \rangle^2 / 2. \tag{7.47}$$

The various procedures under which the OS can be used to define false alarm probabilities and detection probabilities can be found in detail in Refs. (26; 27; 28). Note that the OS estimator was initially developed for the weak signal regime, and can be biased for the intermediate signal regime and beyond. However, its utility can be extended by hybridizing it with Bayesian principles– the noise-marginalized OS (30) averages the OS over noise parameters drawn from an MCMC chain that has been run on a fixed-α PTA GWB search. The OS can be used to explore how the GWB SNR scales with various PTA configuration variables, like timing precision, number of pulsars, GWB amplitude, etc. (31; 32).

It is also possible to take a broader view of the OS as a maximum likelihood estimator, which uses pairwise correlation measurements as the data with heterogeneous uncertainties. In this view, we are free to fit models of inter-pulsar correlation signatures to the measured pairwise correlations as we choose. The pairwise correlations and their uncertainties are (28)

$$\rho_{ab} = \frac{\vec{\delta t}_a^{\text{T}} \boldsymbol{P}_a^{-1} \hat{\boldsymbol{S}}_{ab} \boldsymbol{P}_b^{-1} \vec{\delta t}_b}{\text{Tr} \left[\boldsymbol{P}_a^{-1} \hat{\boldsymbol{S}}_{ab} \boldsymbol{P}_b^{-1} \hat{\boldsymbol{S}}_{ba} \right]},$$

$$\sigma_{ab} = \left(\text{Tr} \left[\boldsymbol{P}_a^{-1} \hat{\boldsymbol{S}}_{ab} \boldsymbol{P}_b^{-1} \hat{\boldsymbol{S}}_{ba} \right] \right)^{-1/2}, \tag{7.48}$$

where now $\boldsymbol{S}_{ab} = A_{\text{GWB}}^2 \Gamma_{ab} \hat{\boldsymbol{S}}_{ab}$, and Γ_{ab} is the ORF or inter-pulsar correlation signature. We can often describe Γ_{ab} as a *linear model*, e.g., in terms of pixel or multipole power for anisotropic modeling (33; 34; 35; 36), in terms of a Legendre or Chebyshev series for an agnostic description (37; 38; 39), or in terms of multiple processes that may involve a SGWB and systematics (40). This simply means that we can write $\Gamma_{ab} = \sum_k c_k X_{ab,k}$, or alternatively in vector notation $\vec{\Gamma} = \boldsymbol{X} \vec{c}$, such that $\boldsymbol{X}$ is a ($N_{\text{pairs}} \times N_{\text{features}}$) design matrix of the basis functions evaluated for the different pulsar pairs. Treating the $\{\rho_{ab}, \sigma_{ab}\}$ in vector form, and assuming σ_{ab} are Gaussian uncertainties, we can use linear regression to deduce maximum-likelihood model estimators with an associated uncertainty covariance matrix:

$$\vec{c}_{\text{ML}} = \left(\boldsymbol{X}^T \boldsymbol{C}^{-1} \boldsymbol{X} \right)^{-1} \boldsymbol{X}^T \boldsymbol{C}^{-1} \vec{\rho}, \quad \Sigma_c = \left(\boldsymbol{X}^T \boldsymbol{C}^{-1} \boldsymbol{X} \right)^{-1}, \tag{7.49}$$

where C is a diagonal matrix of squared σ_{ab} values. The diagonal elements of Σ_c give the variance of the $\vec{c}_{\text{ML}}$ estimates, while the off-diagonal elements describe the parameter covariances.

7.3.1.2 *Bridging the Bayesian Odds Ratio and the Frequentist Optimal statistic*

Under certain circumstances it is possible to relate Bayesian model selection to frequentist hypothesis testing (41). When data is informative such that the likelihood is strongly peaked, the Bayesian evidence can be computed under the *Laplace approximation*, such that

$$Z_{\mathcal{H}} \equiv \int d\theta \, p(d|\theta, \mathcal{H}) p(\theta|\mathcal{H}) \approx p(d|\theta_{\mathrm{ML}}, \mathcal{H}) \Delta V_{\mathcal{H}} / V_{\mathcal{H}}, \qquad (7.50)$$

where $p(d|\theta_{\mathrm{ML}}, \mathcal{H})$ maximizes the likelihood with parameters θ given data d under model $\mathcal{H}$. $\Delta V_{\mathcal{H}}/V_{\mathcal{H}}$ measures the compactness of the parameter space volume occupied by the likelihood with respect to the total prior volume, incorporating the Bayesian notion of model parsimony. Taking the ratio of Bayesian evidences between two models, labeled 1 and 2, and assuming equal prior odds, allows the Bayesian odds ratio, $\mathcal{O}_{12}$, to be written as

$$\ln \mathcal{O}_{12} \approx \ln \Lambda_{\mathrm{ML}}(d) + \ln \left[(\Delta V_1/V_1)/(\Delta V_2/V_2) \right], \qquad (7.51)$$

where $\Lambda_{\mathrm{ML}}(d)$ is the maximum likelihood ratio. The relevant maximum likelihood statistic for PTA GWB detection is the *optimal statistic*, which as we have seen is a noise-weighted two-point correlation statistic between all unique pulsar pairs, comparing models with and without spatial correlations between pulsars (26; 27; 28). From Eq. 7.47, the SNR of such GWB-induced correlations can be written as $\langle \ln \Lambda_{\mathrm{ML}} \rangle = \langle \rho \rangle^2/2$. While the likelihood may be marginally more compact under the model with Hellings & Downs correlations, there is no difference in parameter dimensionality; hence we ignore the likelihood compactness terms. The key relationship between the Bayesian odds ratio in favor of Hellings & Downs correlations and the frequentist SNR of such correlations can then be written as

$$\ln \mathcal{O}_{\mathrm{HD}} \approx \rho^2/2. \qquad (7.52)$$

7.3.2 Individual Binary Statistics

7.3.2.1 $\mathcal{F}_e$ *Statistic*

The $\mathcal{F}_e$ statistic is a maximum likelihood estimator of the sky-location and orbital frequency of an individually-resolvable monochromatic (i.e. circular) SMBBH signal in PTA data. The assumed signal model includes only the *Earth term* of timing delays. This concept was originally developed in the context of detecting continuous GWs from neutron stars (42; 43), then later adapted for PTA continuous GW searches (44; 45). Developing this statistic requires us to re-arrange the signal model of induced timing delays into a format where we can maximize over the coefficients of a set of time-dependent basis functions. We use a parenthetical notation to denote the noise-weighted inner product of two vectors, such that $(\vec{x}|\vec{y}) = \vec{x}^{\mathrm{T}} C^{-1} \vec{y}$, where C is a covariance matrix of noise in the observed vectors $\{\vec{x}, \vec{y}\}$. The log-likelihood ratio between the signal+noise model and the noise-only model can be written as

$$\ln \Lambda = \ln \left(\frac{p(\vec{\delta t}|\vec{s})}{p(\vec{\delta t}|\vec{0})} \right)$$

$$= -\frac{1}{2}(\vec{\delta t} - \vec{s}|\vec{\delta t} - \vec{s}) + \frac{1}{2}(\vec{\delta t}|\vec{\delta t})$$

$$= (\vec{\delta t}|\vec{s}) - \frac{1}{2}(\vec{s}|\vec{s}). \tag{7.53}$$

Using our earlier introduced waveform model for an individually resolvable SMBBH (see Chap. 4), we re-write the Earth-term signal model in terms of coefficients of *extrinsic* variables $(\zeta, \iota, \Phi_0, \psi)$ and basis functions of *intrinsic* variables (θ, ϕ, ω_0) (42)

$$\vec{s}(t, \hat{\Omega}) = \sum_{k=1}^{4} a_k(\zeta, \iota, \Phi_0, \psi) \vec{A}^k(t, \theta, \phi, \omega_0), \tag{7.54}$$

where $\vec{s}$ and $\vec{A}^k$ are concatenated vectors of time-series for all pulsars, such that

$$\vec{s} = \begin{bmatrix} s_1 \\ s_2 \\ \vdots \\ s_{N_p} \end{bmatrix}, \qquad \vec{A}^k = \begin{bmatrix} A_1^k \\ A_2^k \\ \vdots \\ A_{N_p}^k \end{bmatrix}, \tag{7.55}$$

$$A_a^1 = F_a^+(\hat{\Omega})\omega(t)^{-1/2}\sin(2\Phi(t)),$$
$$A_a^2 = F_a^+(\hat{\Omega})\omega(t)^{-1/2}\cos(2\Phi(t)),$$
$$A_a^3 = F_a^\times(\hat{\Omega})\omega(t)^{-1/2}\sin(2\Phi(t)),$$
$$A_a^4 = F_a^\times(\hat{\Omega})\omega(t)^{-1/2}\cos(2\Phi(t)), \tag{7.56}$$

$$a_1 = \zeta[(1 + \cos^2 \iota)\cos 2\Phi_0 \cos 2\psi + 2\cos\iota \sin 2\Phi_0 \sin 2\psi],$$
$$a_2 = -\zeta[(1 + \cos^2 \iota)\sin 2\Phi_0 \cos 2\psi - 2\cos\iota \cos 2\Phi_0 \sin 2\psi],$$
$$a_3 = \zeta[(1 + \cos^2 \iota)\cos 2\Phi_0 \sin 2\psi - 2\cos\iota \sin 2\Phi_0 \cos 2\psi],$$
$$a_4 = -\zeta[(1 + \cos^2 \iota)\sin 2\Phi_0 \sin 2\psi + 2\cos\iota \cos 2\Phi_0 \cos 2\psi]. \tag{7.57}$$

The binary is assumed to be slowly evolving over the course of the pulsar observation baseline, such that $\omega(t) \approx \omega_0$ and $\Phi(t) \approx \omega_0 t$. The log-likelihood can then be written as

$$\ln \Lambda = a_k \mathbf{N}^k - \frac{1}{2}\mathbf{M}^{kl} a_k a_l, \tag{7.58}$$

where $\mathbf{N}^k = (\vec{\delta t}|\vec{A}^k)$ and $\mathbf{M}^{kl} = (\vec{A}^k|\vec{A}^l)$. Summation convention is used to evaluate terms with paired superscript and subscript indices. Maximizing this log-likelihood over the four coefficients, a_k, yields their maximum likelihood estimates

$$a_k^{\mathrm{ML}} = M_{kl} N^l,$$ (7.59)

where $M_{kl} = (M^{kl})^{-1}$. These coefficients can be plugged back into the log-likelihood to yield the $\mathcal{F}_e$ statistic

$$2\mathcal{F}_e = N^k M_{kl} N^l.$$ (7.60)

This statistic can then be mapped over different orbital frequencies and sky locations, or globally maximized to find the best-fit values of these parameters. The form of this statistic's false alarm probability, detection probability, and procedures under which it is used are detailed in Ref. (44; 45). While only the sky location and orbital frequency are explicitly searched over, it is possible to estimate the extrinsic parameters from the maximum likelihood amplitude coefficeints (46; 45). Note also that a generalized version of the $\mathcal{F}_e$-statistic applicable to arbitrary binary eccentricities has also been developed (47), where the space of intrinsic parameters additionally includes binary eccentricity, e, and the initial mean anomaly, $l_0 = 2\pi t_0/P$, where P is the orbital period.

7.3.2.2 $\mathcal{F}_p$ Statistic

As a maximum-likelihood estimator for individually-resolvable monochromatic SMBBH signals in PTA data, the $\mathcal{F}_p$ statistic is similar to the $\mathcal{F}_e$ statistic, except that it also accounts for the *pulsar term* in the GW-induced timing delays. This is sometimes also known as the *incoherent $\mathcal{F}$ statistic* because it involves summing over squares of data quantities, rather than in the $\mathcal{F}_e$ statistic where one squares the sum of data quantities. A central assumption used here when introducing the pulsar term is that the source evolution remains slow enough that the Earth-term orbital frequency, ω_0, and pulsar-term orbital frequency, ω_p, are identical (or at least indistinguishable within the PTA resolution). We can perform a Taylor expansion of the pulsar-term frequency such that

$$\omega(t_p) = \omega_0 \left[1 - \frac{256}{5} \left(\frac{G\mathcal{M}}{c^3} \right)^{5/3} \omega_0^{8/3} t_p \right]^{-3/8},$$

$$\approx \omega_0 \left[1 + \frac{96}{5} \left(\frac{G\mathcal{M}}{c^3} \right)^{5/3} \omega_0^{8/3} (t_e - L_p(1 + \hat{\Omega} \cdot \hat{p})) \right],$$ (7.61)

where $\mathcal{M}$ is the binary chirp mass, L_p is the pulsar distance, $\hat{\Omega} \equiv -(\sin\theta\cos\phi, \sin\theta\sin\phi, \cos\theta)$ is the GW propagation direction for GWs originating from sky-location (θ, ϕ), and $\hat{p}$ is the pulsar unit-vector direction on the sky. The condition that $\omega_p = \omega(t_p) \approx \omega_0$ requires

$$\omega_0 \ll \left[\frac{5}{96} \left(\frac{c^3}{G\mathcal{M}} \right)^{5/3} \left| T - L(1 + \hat{\Omega} \cdot \hat{p}) \right| \right]^{3/8},$$ (7.62)

where T is the total pulsar observation time. With both Earth and pulsar terms sharing a common signal frequency, we can now write the GW signal in pulsar a as

$$s_a(t, \hat{\Omega}) = \sum_{k=1}^{2} b_{k,a}(\zeta, \iota, \Phi_0, \psi, \tilde{\Phi}_a, \theta, \phi) B_a^k(t, \omega_0), \qquad (7.63)$$

where

$$\tilde{\Phi}_a = \omega_0 L_a (1 + \hat{\Omega} \cdot \hat{p}_a)/c + \Phi_0, \qquad (7.64)$$

is the pulsar-term orbital phase. The amplitude and basis functions are defined as

$$
\begin{aligned}
b_{1,a} &= \zeta \left[(1 + \cos^2 \iota)(F_a^+ \cos 2\psi + F_a^\times \sin 2\psi)(\cos 2\Phi_0 - \cos 2\tilde{\Phi}_a) \right. \\
&\quad \left. + 2\cos\iota(F_a^+ \sin 2\psi - F_a^\times \cos 2\psi)(\sin 2\Phi_0 - \sin 2\tilde{\Phi}_a) \right], \\
b_{2,a} &= -\zeta \left[(1 + \cos^2 \iota)(F_a^+ \cos 2\psi + F_a^\times \sin 2\psi)(\sin 2\Phi_0 - \sin 2\tilde{\Phi}_a) \right. \\
&\quad \left. - 2\cos\iota(F_a^+ \sin 2\psi - F_a^\times \cos 2\psi)(\cos 2\Phi_0 - \cos 2\tilde{\Phi}_a) \right], \qquad (7.65)
\end{aligned}
$$

and

$$B_a^1 = \frac{1}{\omega_0^{1/3}} \sin(2\omega_0 t), \quad B_a^2 = \frac{1}{\omega_0^{1/3}} \cos(2\omega_0 t). \qquad (7.66)$$

The log-likelihood ratio can then be written as

$$\ln \Lambda = \sum_{a=1}^{N_p} \left[b_{k,a} P_a^k - \frac{1}{2} Q^{kl} b_{k,a} b_{l,a} \right], \qquad (7.67)$$

where $P_a^k = (\delta t_a | B_a^k)$ and $Q^{kl} = (B_a^k | B_a^l)$. Maximizing $\ln \Lambda$ over the $2N_p$ coefficients yields

$$b_{k,a}^{\text{ML}} = (Q_a^{kl})^{-1} P_a^k, \qquad (7.68)$$

which, when inserted back into the log-likelihood ratio, give

$$2\mathcal{F}_p = \sum_{a=1}^{N_p} P_a^k (Q_a^{kl})^{-1} P_a^l. \qquad (7.69)$$

This statistic is more often used than the $\mathcal{F}_e$ statistic because of the realism of the model that includes the pulsar term, and all factors involving the (uncertain) pulsar distances have been absorbed into amplitude coefficients that are maximized over. It can be used to examine pulsar-timing data to compute a value of the $\mathcal{F}_p$ statistic as a function of frequency, and in so doing produce sensitivity curves for indivudally-resolvable binary GW sources. The form of this statistic's false alarm probability, detection probability, and procedures under which it is used are detailed in Ref. (45). Note also that an alternative to the $\mathcal{F}_p$ statistic's maximization over amplitude coefficients is to marginalize over these, or to maximize over the physical pulsar-term phase parameters, $\tilde{\Phi}_a$. Both schemes were explored in Ref. (48) with the former marginalization technique referred to as the $\mathcal{B}_p$ statistic.

Bibliography

[1] Rutger van Haasteren and Michele Vallisneri. New advances in the Gaussian-process approach to pulsar-timing data analysis. *Physical Review D*, 90(10):104012, November 2014. 7, 7.1.1, 5, 7.2.1

[2] L Lentati, MP Hobson, and P Alexander. Bayesian estimation of non-Gaussianity in pulsar timing analysis. *Monthly Notices of the Royal Astronomical Society*, 444(4):3863–3878, November 2014. 1

[3] JA Ellis and NJ Cornish. Transdimensional Bayesian approach to pulsar timing noise analysis. *Physical Review D*, 93(8):084048, April 2016. 1

[4] Ryan M Shannon and James M Cordes. Assessing the role of spin noise in the precision timing of millisecond pulsars. *The Astrophysical Journal*, 725(2):1607–1619, December 2010. 7.1.2

[5] Rutger van Haasteren and Michele Vallisneri. Low-rank approximations for large stationary covariance matrices, as used in the Bayesian and generalized-least-squares analysis of pulsar-timing data. *Monthly Notices of the Royal Astronomical Society*, 446(2):1170–1174, January 2015. 7.1.2

[6] L Sampson, NJ Cornish, and ST McWilliams. Constraining the solution to the last parsec problem with pulsar timing. *Physical Review D*, 91(8):084055, April 2015. 7.1.2

[7] Alberto Sesana, Alberto Vecchio, and C N Colacino. The stochastic gravitational-wave background from massive black hole binary systems: implications for observations with Pulsar Timing Arrays. *Monthly Notices of the Royal Astronomical Society*, 390(1):192–209, October 2008. 7.1.2

[8] Siyuan Chen, Hannah Middleton, Alberto Sesana, et al. Probing the assembly history and dynamical evolution of massive black hole binaries with pulsar timing arrays. *Monthly Notices of the Royal Astronomical Society*, 468(1):404–417, June 2017. 7.1.2

[9] Siyuan Chen, Alberto Sesana, and Walter Del Pozzo. Efficient computation of the gravitational wave spectrum emitted by eccentric massive black hole binaries in stellar environments. *Monthly Notices of the Royal Astronomical Society*, 470(2):1738–1749, September 2017. 7.1.2

[10] Siyuan Chen, Alberto Sesana, and Christopher J Conselice. Constraining astrophysical observables of galaxy and supermassive black hole binary mergers using pulsar timing arrays. *Monthly Notices of the Royal Astronomical Society*, 488(1):401–418, September 2019. 7.1.2

[11] Stephen R Taylor, Joseph Simon, and Laura Sampson. Constraints on the dynamical environments of supermassive black-hole binaries using pulsar-timing arrays. *Physical Review Letters*, 118(18):181102, May 2017. 7.1.2

[12] K Aggarwal, Z Arzoumanian, PT Baker, et al. The NANOGrav 11 yr data set: limits on gravitational waves from individual supermassive black hole binaries. *The Astrophysical Journal*, 880(2):116, August 2019. 7.1.2

[13] Lindley Lentati, P Alexander, MP Hobson, et al. Hyper-efficient model-independent Bayesian method for the analysis of pulsar timing data. *Physical Review D*, 87(10):104021, May 2013. 7.1.2, 7.2.1, 7.2.2

[14] KJ Lee, CG Bassa, GH Janssen, et al. Model-based asymptotically optimal dispersion measure correction for pulsar timing. *Monthly Notices of the Royal Astronomical Society*, 441(4):2831–2844, July 2014. 7.1.3

[15] MJ Keith, W Coles, RM Shannon, et al. Measurement and correction of variations in interstellar dispersion in high-precision pulsar timing. *Monthly Notices of the Royal Astronomical Society*, 429(3):2161–2174, March 2013. 7.1.3

[16] JM Cordes and RM Shannon. A measurement model for precision pulsar timing. *arXiv e-prints*, page arXiv:1010.3785, October 2010. 7.1.3

[17] L Lentati, RM Shannon, WA Coles, et al. From spin noise to systematics: stochastic processes in the first International Pulsar Timing Array data release. *Monthly Notices of the Royal Astronomical Society*, 458(2):2161–2187, May 2016. 7.1.4

[18] Rutger van Haasteren and Yuri Levin. Understanding and analysing time-correlated stochastic signals in pulsar timing. *Monthly Notices of the Royal Astronomical Society*, 428(2):1147–1159, January 2013. 5, 7.2.2

[19] Jack Sherman and Winifred J Morrison. Adjustment of an inverse matrix corresponding to a change in one element of a given matrix. *Ann. Math. Statist.*, 21(1):124–127, 03 1950. 7.2.1

[20] Radford M. Neal. Slice sampling. *The Annals of Statistics*, 31(3):705–741, 2003. 7.2.1

[21] Michele Vallisneri and Rutger van Haasteren. Taming outliers in pulsar-timing data sets with hierarchical likelihoods and Hamiltonian sampling. *Monthly Notices of the Royal Astronomical Society*, 466(4):4954–4959, April 2017. 7.2.1

[22] Max A Woodbury. Inverting modified matrices. *Memorandum report*, 42(106):336, 1950. 7.2.2

[23] Cholesky, André-Louis. "On the Numerical Solving of Systems of Linear Equations." *Sabix Bulletin. Society of Friends of the Library and History of École Polytechnique* 39 (2005): 81-95. 7.2.2

[24] Norbert Wiener et al. Generalized harmonic analysis. *Acta mathematica*, 55:117–258, 1930. 7.2.2

[25] Alexander Khintchine. Correlation theory of the station "a ren stochastic processes. *mathematical annals*, 109(1):604–615, 1934. 7.2.2

[26] Melissa Anholm, Stefan Ballmer, Jolien DE Creighton, et al. Optimal strategies for gravitational wave stochastic background searches in pulsar timing data. *Physical Review D*, 79(8):084030, April 2009. 7.3.1.1, 7.3.1.1, 7.3.1.2

[27] PB Demorest, RD Ferdman, ME Gonzalez, et al. Limits on the Stochastic Gravitational Wave Background from the North American Nanohertz Observatory for Gravitational Waves. *The Astrophysical Journal*, 762(2):94, January 2013. 7.3.1.1, 7.3.1.1, 7.3.1.2

[28] Sydney J Chamberlin, Jolien DE Creighton, Xavier Siemens, et al. Time-domain implementation of the optimal cross-correlation statistic for stochastic gravitational-wave background searches in pulsar timing data. *Physical Review D*, 91(4):044048, February 2015. 7.3.1.1, 7.3.1.1, 7.3.1.2

[29] JA Ellis, X Siemens, and R van Haasteren. An efficient approximation to the likelihood for gravitational wave stochastic background detection using pulsar timing data. *The Astrophysical Journal*, 769(1):63, May 2013. 7.3.1.1, 7.3.1.1

[30] Sarah J Vigeland, Kristina Islo, Stephen R Taylor, and Justin A Ellis. Noise-marginalized optimal statistic: A robust hybrid frequentist-Bayesian statistic for the stochastic gravitational-wave background in pulsar timing arrays. *Physical Review D*, 98(4):044003, August 2018. 7.3.1.1

[31] Xavier Siemens, Justin Ellis, Fredrick Jenet, and Joseph D Romano. The stochastic background: scaling laws and time to detection for pulsar timing arrays. *Classical and Quantum Gravity*, 30(22):224015, November 2013. 7.3.1.1

[32] SJ Vigeland and X Siemens. Supermassive black hole binary environments: Effects on the scaling laws and time to detection for the stochastic background. *Physical Review D*, 94(12):123003, December 2016. 7.3.1.1

[33] Eric Thrane, Stefan Ballmer, Joseph D Romano, et al. Probing the anisotropies of a stochastic gravitational-wave background using a network of ground-based laser interferometers. *Physical Review D*, 80(12):122002, December 2009. 7.3.1.1

[34] Stephen R Taylor and Jonathan R Gair. Searching for anisotropic gravitational-wave backgrounds using pulsar timing arrays. *Physical Review D*, 88(8):084001, October 2013. 7.3.1.1

[35] CMF Mingarelli, T Sidery, I Mandel, and A Vecchio. Characterizing gravitational wave stochastic background anisotropy with pulsar timing arrays. *Physical Review D*, 88(6):062005, September 2013. 7.3.1.1

[36] Stephen R Taylor, Rutger van Haasteren, and Alberto Sesana. From bright binaries to bumpy backgrounds: Mapping realistic gravitational wave skies with pulsar-timing arrays. *Physical Review D*, 102(8):084039, October 2020. 7.3.1.1

[37] Jonathan Gair, Joseph D Romano, Stephen Taylor, and Chiara MF Mingarelli. Mapping gravitational-wave backgrounds using methods from CMB analysis: Application to pulsar timing arrays. *Physical Review D*, 90(8):082001, October 2014. 7.3.1.1

[38] L Lentati, SR Taylor, CMF Mingarelli, et al. European Pulsar Timing Array limits on an isotropic stochastic gravitational-wave background. *Monthly Notices of the Royal Astronomical Society*, 453:2576–2598, November 2015. 7.3.1.1

[39] Elinore Roebber. Ephemeris Errors and the Gravitational-wave Signal: Harmonic Mode Coupling in Pulsar Timing Array Searches. *The Astrophysical Journal*, 876(1):55, May 2019. 7.3.1.1

[40] C Tiburzi, G Hobbs, M Kerr, et al. A study of spatial correlations in pulsar timing array data. *Monthly Notices of the Royal Astronomical Society*, 455(4):4339–4350, February 2016. 7.3.1.1

[41] Joseph D Romano and Neil. J Cornish. Detection methods for stochastic gravitational-wave backgrounds: a unified treatment. *Living Reviews in Relativity*, 20(1):2, April 2017. 7.3.1.2

[42] Piotr Jaranowski, Andrzej Królak, and Bernard F. Schutz. Data analysis of gravitational-wave signals from spinning neutron stars: The signal and its detection. *Physical Review D*, 58(6):063001, September 1998. 7.3.2.1, 7.3.2.1

[43] Curt Cutler and Bernard F Schutz. Generalized F-statistic: Multiple detectors and multiple gravitational wave pulsars. *Physical Review D*, 72(6):063006, September 2005. 7.3.2.1

[44] Stanislav Babak and Alberto Sesana. Resolving multiple supermassive black hole binaries with pulsar timing arrays. *Physical Review D*, 85(4):044034, February 2012. 7.3.2.1, 7.3.2.1

[45] JA Ellis, X Siemens, and JDE Creighton. Optimal Strategies for Continuous Gravitational Wave Detection in Pulsar Timing Arrays. *The Astrophysical Journal*, 756(2):175, September 2012. 7.3.2.1, 7.3.2.1, 7.3.2.2

[46] Neil J Cornish and Edward K Porter. The search for massive black hole binaries with LISA. *Classical and Quantum Gravity*, 24(23):5729–5755, December 2007. 7.3.2.1

[47] SR Taylor, EA Huerta, JR Gair, and ST McWilliams, Detecting eccentric supermassive black hole binaries with pulsar timing arrays: resolvable source strategies. *The Astrophysical Journal*, 817(1):70, January 2016. 7.3.2.1

[48] Stephen Taylor, Justin Ellis, and Jonathan Gair. Accelerated Bayesian model-selection and parameter-estimation in continuous gravitational-wave searches with pulsar-timing arrays. *Physical Review D*, 90(10):104028, November 2014. 7.3.2.2

CHAPTER 8

The Past, Present, and Future of PTAs

By the very nature of the signals they hunt, PTAs are long timescale experiments. The field has taken some time to gain traction, with the slow trickle in accumulated information being mirrored by a steady growth in interest and activity of the regional collaborations. There are three major regional PTA collaborations that each have greater than a decade of precision timing observations, all inaugurated at roughly the same time in the early 2000s. Ironically, at the same time as the regional PTAs formed, the most constraining limit on the SGWB came from the combination of international datasets recorded in Australia (Parkes Radio Telescope) and Puerto Rico (Arecibo Radio Telescope) (1), corresponding to 1.1×10^{-14} at $f = 1/\text{yr}^{-1}$ with 95% confidence for a SGWB with $\alpha = -2/3$. While earlier campaigns had produced limits before this (2; 3; 4; 5; 6),[1] Ref. (1) was the first to be within an order of magnitude of current results. The limits quoted below will all correspond to the characteristic strain at $f = 1/\text{yr}^{-1}$ for a power-law with index $-2/3$, either with 95% confidence for frequentist studies or 95% credibility for Bayesian studies.

The Parkes Pulsar Timing Array (PPTA) uses the 64 m Parkes Radio Telescope located in Parkes, New South Wales, Australia to time 26 millisecond pulsars over a current baseline of ~17 years (since 2004) (7). The PPTA is distinguished amongst other PTA collaborations through its access to the southern hemisphere, allowing it to observe one of the highest quality pulsars, PSR J0437−4715 (although see Ref. (8; 9)). Throughout its operations, the PPTA has boasted some of the tightest constraints on the amplitude of the SGWB (e.g., 10); its most recent published 95% upper limit was 10^{-15} at $f = 1/\text{yr}^{-1}$ (11). However, this limit used only four pulsars and an outdated Solar-system ephemeris model (DE421, released in 2008 (12)). Several

---

[1]In fact, Ref. (5) from 1996 was the first to suggest a Bayesian approach.

DOI: 10.1201/9781003240648-8

analyses have shown that relying on a small number of pulsars (13) and older ephemerides (14; 15) can lead to systematic biases in SGWB constraints. The PPTA is currently analyzing its second data release (7).

The European Pulsar Timing Array (EPTA) is a collaboration using five radio telescopes spread across several countries: the 94 m-equivalent Westerbork Synthesis Radio Telescope (Westerbork, Netherlands), the 100 m Effelsberg Radio Telescope (Bad Münstereifel, Germany), the 76 m Lovell Telescope (Jodrell Bank, UK), the 94 m Nançay Radio Telescope (Sologne, France), and the 64 m Sardinia Radio Telescope (San Basilio, Sardinia, Italy). Established in 2006, the EPTA was the first PTA collaboration to use Bayesian statistical inference to simultaneously model the intrinsic noise in each pulsar, the cross correlations between all pulsars, and the amplitude and spectral index of the SGWB's strain spectrum (16), finding a 95% limit of 6×10^{-15} from five pulsars. It incorporates some of the longest observed pulsars in its datasets; in its most recent data release in 2016, it included 42 pulsars whose baselines ranged from $\sim 7 - 18$ years until 2014 (17), of which six pulsars were used to place a limit of 3×10^{-15} (18). The collaboration is currently working on a new data release.

The North American Nanohertz Observatory for Gravitational waves (NANOGrav), which was established in 2007, uses the 100 m Green Bank Telescope (Green Bank, West Virginia, USA) and the (former) 305 m Arecibo Observatory (Arecibo, Puerto Rico, USA) to time ~ 77 pulsars over a baseline as long as ~ 15 years (19). The collaboration has steadily added new pulsars to its array, where the number of analyzed pulsars in its most recent SGWB searches increased from 17 in its 5-year dataset (limit of 7×10^{-15}) (20), 18 in its 9-year dataset (limit of 1.5×10^{-15}) (21), 34 in its 11-year dataset (limit of 1.45×10^{-15} (14), but see also Ref. (22)), and 45 in its 12.5 year dataset (detection of a common process with median amplitude 1.92×10^{-15}) (23). Notice the trend in the last few datasets: the limit appeared to decrease rapidly, saturate, then increase as a detection of a low-frequency common process was made. The saturation occurred because the NANOGrav array hit a systematic noise floor corresponding to Solar-system ephemeris precision, leading to the development of a Bayesian ephemeris model to mitigate this. Revision of the priors for intrinsic pulsar red noise in the 11-year analysis has shown greater consistency with the 12.5 year results (22). With regards to the latter, NANOGrav has now entered a regime of overwhelming statistical evidence for a common-spectrum low-frequency process being positively supported by 10 pulsars amongst the 45 analyzed. This process has median amplitude 1.92×10^{-15} and $5\% - 95\%$ quantiles of $1.37 - 2.67 \times 10^{-15}$. The Bayes factor in favor of this common-spectrum process versus merely intrinsic per-pulsar noise is in excess of $10^5 : 1$. While this may potentially be the first signs of the SGWB emerging from noise (24), no evidence has yet been found for the distinctive Hellings & Downs inter-pulsar correlation signature from an isotropic background of GWs. The collaboration is currently working on a ~15 year dataset with the number of pulsars exceeding 60.

What was once steady growth and advancement now feels like a torrent of progress. Emerging PTA collaborations such as the Indian PTA (InPTA) (25), the Chinese PTA (CPTA) (26), and more telescope-centered groups like CHIME (27) and MeerTime (28), are ramping up efforts to join the hunt. In fact, the InPTA has recently joined the International Pulsar Timing Array (IPTA), which was originally founded as a consortium of the three major PTA consortia (NANOGrav, EPTA, PPTA). The IPTA has produced two data releases so far, which usually lag several years behind when the corresponding components are released by regional PTAs. There is good reason for this; data combination is a delicate and challenging process, and where pulsars are observed in common, this requires timing solutions and noise models to be harmonized across different telescopes and processing pipelines. The first IPTA data release (29) consisted of 49 pulsars, and was built from NANOGrav's 5 year dataset (20), the extended first PPTA data release (30; 31), the first EPTA data release (17), and publicly available data from Kaspi *et al.* (1994) (4) on PSRs J1857+0943 and J1939+2134 and from Zhu *et al.* 2015 (32) on PSR J1713+0747. The most sensitive four pulsars were used to derive an upper limit of 1.7×10^{-15}. The second IPTA data release (33), consisting of 65 pulsars, adds to the first data release through expanding the NANOGrav component to the 9 year data release (34) and adding additional PPTA data that was included in Shannon *et al.* (2015) (11). Work is ongoing to perform a full-array SGWB search within this dataset.

So what's next? At least one PTA collaboration has detected a common-spectrum process that could be the first trumpet announcing a cresting SGWB detection. This must be confirmed (or refuted) by other regional PTAs and the IPTA. As of writing, these are all in the works. If this is an early announcement of the SGWB, then the amplitude is a little larger than recent models have suggested, although well within the prediction spread (35). Detection of the Hellings & Downs inter-pulsar correlations should follow with only a few more years of data beyond existing baselines (36; 37; 13; 38; 39), and will come packaged with $\sim 40\%$ constraints on the amplitude and spectral index of the characteristic strain spectrum (39). This precision should be sufficient to test some astrophysical models of the SGWB and discriminate the origin under simple assumptions of the spectral index. After that, attention will shift to inference of spectral features such as turnovers that can indicate continued environmental coupling of the SMBHB population into the nanohertz band (40; 41; 42).

Beyond spectral characterization, the inter-pulsar correlations encode a map of the GW sky, allowing PTAs to probe the angular structure of the SGWB signal and potentially unveil the first departures from statistical isotropy (43; 44; 45; 46). Bright pixels or extended regions of angular power could be the first indicators of nearby or massive single sources to which the PTA is slowly becoming sensitive enough to individually resolve (46). This brings us to the next big milestone for PTAs, and a goal that is being actively searched for with only slightly less fervor than the SGWB– individual

SMBHB signals. While we think the SGWB will be detected first (37), resolution and characterization of individual SMBHB systems will confirm once and for all that they are the dominant source class for PTAs, and dangle the tantalizing opportunity for multi-messenger detections of massive black-hole binaries years before LISA will fly (e.g., 47). While much of this chapter has been focused on ever-improving constraints on the SGWB, the strides in single source constraints have been huge; NANOGrav has placed constraints on the strain amplitude corresponding to $h \leq 7.3 \times 10^{-15}$ at $f = 8$ nHz with its 11 year dataset, and in its most sensitive sky location has ruled out $\mathcal{M} = 10^9 (10^{10}) M_\odot$ binaries closer than 120 Mpc (5.5 Gpc) with 95% credibility (48). Recent constraints from the EPTA (49) and PPTA (50) are broadly similar. NANOGrav has also used its 11-year dataset to improve constraints on a prominent binary candidate, 3C66B, finding that guiding constraints on the period of the fiducial binary that are less than an order of magnitude can in turn improve the chirp mass constraints by an order of magnitude (51). Furthermore, NANOGrav recently compiled a target list of 216 massive galaxies within its sensitivity volume, placing multimessenger constraints on the secondary black-hole mass of a fiducial binary, and finding that 19 systems could be constrained to within the same precision as SMBHB constraints in our own Milky Way (52). These results demonstrate the power of the multimessenger goal, and the excitement for the future. PTAs may be able to detect several individual binaries by the end of the 2020s (37; 53; 54), and if any of these have detectable lightcurve variability from accretion (55; 56; 57; 58) or Doppler-boosting variability (59; 60), then the Rubin Observatory's Legacy Survey of Space and Time could snag the electromagnetic counterpart (47).

Speculation further into the future runs into many unknowns, but the goals are clear. We aim to perform precision constraints on modified gravity theories through the presence of alternative GW polarization modes that may be sub-luminal; these modes will have their own distinct overlap reduction functions and lead to departures from the Hellings & Downs curve. Beyond SMBHBs (and associated background, continuous, and burst searches), significant effort must be devoted to spectral separation of the binary signal from cosmological signals lurking beneath. These could include a primordial GW background, cosmic string signatures, imprints of a first-order cosmological phase transition, or heralds of unconstrained fundamental physics, such as the nature of dark matter. Addressing all of these important questions will require sophisticated physical models, robust statistical techniques, and powerful radio facilities to carry us into the middle of this century and beyond. The recent loss of Arecibo is a major blow, but NANOGrav is planning to shift most observations over to the Green Bank Telescope, and legacy Arecibo observations will continue to have impactful weight on future GW searches. In the longer term, the US community requires a replacement world-class radio facility (such as the DSA-2000 concept (61) and/or the ngVLA (62)). The successive stages of the Square Kilometre Array (63) in South Africa and Australia will be transformative for PTA science (64), providing a major boost to the number of pulsars and TOA precision. In China, beyond pulsar surveys and timing with the Five-hundred-meter Aperture Spherical Telescope (FAST) (65; 66), construction is underway on the 110 m Xingjiang QTT (67) and the 120 m Jingdong Radio Telescope.

Pulsar-timing campaigns that focus on searching for gravitational waves are a mature enterprise. Though with exciting possibilities on the horizon, it feels as if our work is just beginning. There is much to discover, resolve, and strategize in order to steer our sensitivity beyond current possibilities toward future opportunities (68; 69). PTAs offer a radically different means to GW detection, wherein a collection of dense, compact astrophysical objects are themselves an integral component of our detector. Much of this book has focused on GWs to the neglect of the extraordinary science of pulsars and the ionized interstellar medium; justice is given to those topics in other volumes by better-qualified authors. For now, I am invigorated by the state of this field and what may soon be possible. I invite the casual reader to join us on this hunt, and hope the expert reader finds this has been a useful overview of the state of the art of nanohertz GW searches with PTAs.

Bibliography

[1] FA Jenet, GB Hobbs, W van Straten, et al. Upper bounds on the low-frequency stochastic gravitational wave background from pulsar timing observations: Current limits and future prospects. *The Astrophysical Journal*, 653:1571–1576, December 2006. 8

[2] RW Hellings and GS Downs. Upper limits on the isotropic gravitational radiation background from pulsar timing analysis. *The Astrophysical Journal*, 265:L39–L42, February 1983. 8

[3] DR Stinebring, MF Ryba, JH Taylor, and RW Romani. Cosmic gravitational-wave background: Limits from millisecond pulsar timing. *Physical Review Letters*, 65(3):285–288, July 1990. 8

[4] VM Kaspi, JH Taylor, and MF Ryba. High-precision timing of millisecond pulsars. III. Long-term monitoring of PSRs B1855+09 and B1937+21. *The Astrophysical Journal*, 428:713, June 1994. 8

[5] MP McHugh, G Zalamansky, F Vernotte, and E Lantz. Pulsar timing and the upper limits on a gravitational wave background: A Bayesian approach. *Physical Review D*, 54(10):5993–6000, November 1996. 8, 1

[6] AN Lommen. New limits on gravitational radiation using pulsars. In W Becker, H Lesch, and J Trümper, editors, *Neutron Stars, Pulsars, and Supernova Remnants*, page 114, January 2002. 8

[7] Matthew Kerr, Daniel J Reardon, George Hobbs, et al. The Parkes pulsar timing array project: Second data release. *Publications of the Astronomical Society of Australia*, 37:e020, June 2020. 8

[8] MT Lam and JS Hazboun. Precision timing of PSR J0437-4715 with the IAR observatory and implications for low-frequency gravitational wave source sensitivity. *The Astrophysical Journal*, 911(2):137, April 2021. ↖

[9] V Sosa Fiscella, S del Palacio, L Combi, et al. PSR J0437-4715: The Argentine Institute of Radioastronomy 2019-2020 Observational Campaign. *The Astrophysical Journal*, 908(2):158, February 2021. ↖

[10] RM Shannon, V Ravi, WA Coles, et al. Gravitational-wave limits from pulsar timing constrain supermassive black hole evolution. *Science*, 342:334–337, October 2013. ↖

[11] RM Shannon, V Ravi, LT Lentati, et al. Gravitational waves from binary supermassive black holes missing in pulsar observations. *Science*, 349(6255):1522–1525, September 2015. ↖

[12] William M Folkner, James G Williams, and Dale H Boggs. The planetary and lunar ephemeris de 421. *IPN progress report*, 42(178):1–34, 2009. ↖

[13] SR Taylor, M Vallisneri, JA Ellis, et al. Are we there yet? time to detection of nanohertz gravitational waves based on pulsar-timing array limits. *The Astrophysical Journal*, 819(1):L6, March 2016. ↖

[14] Z Arzoumanian, PT Baker, A Brazier, et al. The NANOGrav 11 Year data set: pulsar-timing constraints on the stochastic gravitational-wave background. *The Astrophysical Journal*, 859(1):47, May 2018. ↖

[15] M Vallisneri, SR Taylor, J Simon, et al. Modeling the uncertainties of solar system ephemerides for robust gravitational-wave searches with pulsar-timing arrays. *The Astrophysical Journal*, 893(2):112, April 2020. ↖

[16] R van Haasteren, Y Levin, GH Janssen, et al. Placing limits on the stochastic gravitational-wave background using European Pulsar Timing Array data. *Monthly Notices of the Royal Astronomical Society*, 414(4):3117–3128, July 2011. ↖

[17] G Desvignes, RN Caballero, L. Lentati, et al. High-precision timing of 42 millisecond pulsars with the European Pulsar Timing Array. *Monthly Notices of the Royal Astronomical Society*, 458:3341–3380, May 2016. ↖

[18] L Lentati, SR Taylor, CMF Mingarelli, et al. European Pulsar Timing Array limits on an isotropic stochastic gravitational-wave background. *Monthly Notices of the Royal Astronomical Society*, 453:2576–2598, November 2015. ↖

[19] Scott Ransom, A Brazier, S Chatterjee, et al. The NANOGrav program for gravitational waves and fundamental physics. In *Bulletin of the American Astronomical Society*, volume 51, page 195, September 2019. ↖

[20] PB Demorest, RD Ferdman, ME Gonzalez, et al. Limits on the stochastic gravitational wave background from the north american nanohertz observatory for gravitational waves. *The Astrophysical Journal*, 762(2):94, January 2013.

[21] Z Arzoumanian, A Brazier, S Burke-Spolaor, et al. The NANOGrav nine-year data set: Limits on the isotropic stochastic gravitational wave background. *The Astrophysical Journal*, 821:13, April 2016.

[22] Jeffrey S Hazboun, Joseph Simon, Xavier Siemens, and Joseph D Romano. Model dependence of bayesian gravitational-wave background statistics for pulsar timing arrays. *The Astrophysical Journal*, 905(1):L6, December 2020.

[23] Zaven Arzoumanian, Paul T Baker, Harsha Blumer, et al. The NANOGrav 12.5 yr data set: search for an isotropic stochastic gravitational-wave background. *The Astrophysical Journal*, 905(2):L34, December 2020.

[24] Joseph D Romano, Jeffrey S Hazboun, Xavier Siemens, and Anne M Archibald. Common-spectrum process versus cross-correlation for gravitational-wave searches using pulsar timing arrays. *Physical Review D*, 103(6):063027, March 2021.

[25] Bhal Chandra Joshi, Prakash Arumugasamy, Manjari Bagchi, et al. Precision pulsar timing with the ort and the gmrt and its applications in pulsar astrophysics. *Journal of Astrophysics and Astronomy*, 39(4):1–10, 2018.

[26] KJ Lee. Prospects of gravitational wave detection using pulsar timing array for chinese future telescopes. In *Frontiers in Radio Astronomy and FAST Early Sciences Symposium 2015*, volume 502, page 19, 2016.

[27] Cherry Ng. Pulsar science with the CHIME telescope. In P Weltevrede, BBP Perera, LL Preston, and S. Sanidas, editors, *Pulsar Astrophysics the Next Fifty Years*, volume 337, pages 179–182, August 2018.

[28] M Bailes, E Barr, NDR Bhat, et al. MeerTime - the MeerKAT key science program on pulsar timing. In *MeerKAT Science: On the Pathway to the SKA*, page 11, January 2016.

[29] JPW Verbiest, L Lentati, G Hobbs, et al. The international pulsar timing array: First data release. *Monthly Notices of the Royal Astronomical Society*, 458(2):1267–1288, May 2016.

[30] RN Manchester, G Hobbs, M Bailes, et al. The parkes pulsar timing array project. *Publications of the Astronomical Society of Australia*, 30:e017, January 2013.

[31] DJ Reardon, G Hobbs, W Coles, et al. Timing analysis for 20 millisecond pulsars in the Parkes Pulsar Timing Array. *Monthly Notices of the Royal Astronomical Society*, 455(2):1751–1769, January 2016.

[32] WW Zhu, IH Stairs, PB Demorest, et al. Testing theories of gravitation using 21-year timing of pulsar binary J1713+0747. *The Astrophysical Journal*, 809(1):41, August 2015.

[33] BBP Perera, ME DeCesar, PB Demorest, et al. The International pulsar timing array: Second data release. *Monthly Notices of the Royal Astronomical Society*, 490(4):4666–4687, December 2019.

[34] NANOGrav Collaboration, Zaven Arzoumanian, Adam Brazier, et al. The NANOGrav nine-year data set: Observations, arrival time measurements, and analysis of 37 millisecond pulsars. *The Astrophysical Journal*, 813(1):65, November 2015.

[35] H Middleton, A Sesana, S Chen, et al. Massive black hole binary systems and the NANOGrav 12.5 yr results. *Monthly Notices of the Royal Astronomical Society*, 502(1):L99–L103, March 2021.

[36] Xavier Siemens, Justin Ellis, Fredrick Jenet, and Joseph D. Romano. The stochastic background: Scaling laws and time to detection for pulsar timing arrays. *Classical and Quantum Gravity*, 30(22):224015, November 2013.

[37] Pablo A Rosado, Alberto Sesana, and Jonathan Gair. Expected properties of the first gravitational wave signal detected with pulsar timing arrays. *Monthly Notices of the Royal Astronomical Society*, 451(3):2417–2433, August 2015.

[38] Luke Zoltan Kelley, Laura Blecha, Lars Hernquist, et al. The Gravitational wave background from massive black hole binaries in illustris: Spectral features and time to detection with pulsar timing arrays. *arXiv.org*, page arXiv:1702.02180, February 2017.

[39] Nihan S Pol, Stephen R Taylor, Luke Zoltan Kelley, et al. Astrophysics milestones for pulsar timing array gravitational-wave detection. *The Astrophysical Journal*, 911(2):L34, April 2021.

[40] L Sampson, NJ Cornish, and ST McWilliams. Constraining the solution to the last parsec problem with pulsar timing. *Physical Review D*, 91(8):084055, April 2015.

[41] SR Taylor, J Simon, and L Sampson. Constraints on the dynamical environments of supermassive black-hole binaries using pulsar-timing arrays. *Physical Review Letters*, 118(18):181102, May 2017.

[42] Siyuan Chen, Alberto Sesana, and Christopher J Conselice. Constraining astrophysical observables of galaxy and supermassive black hole binary mergers using pulsar timing arrays. *Monthly Notices of the Royal Astronomical Society*, 488(1):401–418, September 2019. 8

[43] CMF Mingarelli, T Sidery, I Mandel, and A Vecchio. Characterizing gravitational wave stochastic background anisotropy with pulsar timing arrays. *Physical Review D*, 88(6):062005, September 2013. 8

[44] Stephen R Taylor and Jonathan R Gair. Searching for anisotropic gravitational-wave backgrounds using pulsar timing arrays. *Physical Review D*, 88(8):084001, October 2013. 8

[45] SR Taylor, CMF Mingarelli, JR Gair, et al. Limits on anisotropy in the nanohertz stochastic gravitational wave background. *Physical Review Letters*, 115(4):041101, July 2015. 8

[46] Stephen R Taylor, Rutger van Haasteren, and Alberto Sesana. From bright binaries to bumpy backgrounds: Mapping realistic gravitational wave skies with pulsar-timing arrays. *Physical Review D*, 102(8):084039, October 2020. 8

[47] Luke Zoltan Kelley, Zoltán Haiman, Alberto Sesana, and Lars Hernquist. Massive BH binaries as periodically variable AGN. *Monthly Notices of the Royal Astronomical Society*, 485(2):1579–1594, May 2019. 8

[48] K Aggarwal, Z Arzoumanian, PT Baker, et al. The NANOGrav 11 yr data set: Limits on gravitational waves from individual supermassive black hole binaries. *The Astrophysical Journal*, 880(2):116, August 2019. 8

[49] S Babak, A Petiteau, A Sesana, et al. European Pulsar Timing Array limits on continuous gravitational waves from individual supermassive black hole binaries. *Monthly Notices of the Royal Astronomical Society*, 455:1665–1679, January 2016. 8

[50] XJ Zhu, G Hobbs, L Wen, et al. An all-sky search for continuous gravitational waves in the Parkes Pulsar Timing Array data set. *Monthly Notices of the Royal Astronomical Society*, 444(4):3709–3720, November 2014. 8

[51] Zaven Arzoumanian, Paul T Baker, Adam Brazier, et al. Multimessenger gravitational-wave searches with pulsar timing arrays: Application to 3C 66B Using the NANOGrav 11-year Data Set. *The Astrophysical Journal*, 900(2):102, September 2020. 8

[52] Zaven Arzoumanian, Paul T Baker, Adam Brazier, et al. The NANOGrav 11yr data set: Limits on supermassive black hole binaries in galaxies within 500Mpc. *The Astrophysical Journal*, 914(2):121, 2021. 8

[53] Chiara MF Mingarelli, T Joseph W Lazio, Alberto Sesana, et al. The local nanohertz gravitational-wave landscape from supermassive black hole binaries. *Nature Astronomy*, 1:886–892, November 2017. 8

[54] Luke Zoltan Kelley, Laura Blecha, Lars Hernquist, et al. Single sources in the low-frequency gravitational wave sky: properties and time to detection by pulsar timing arrays. *Monthly Notices of the Royal Astronomical Society*, 477(1):964–976, June 2018. 8

[55] Brian D Farris, Paul Duffell, Andrew I MacFadyen, and Zoltan Haiman. Binary Black Hole Accretion from a Circumbinary Disk: Gas Dynamics inside the Central Cavity. *The Astrophysical Journal*, 783(2):134, March 2014. 8

[56] Daniel J D'Orazio, Zoltán Haiman, and Andrew MacFadyen. Accretion into the central cavity of a circumbinary disc. *Monthly Notices of the Royal Astronomical Society*, 436(4):2997–3020, December 2013. 8

[57] Paul C Duffell, Daniel D'Orazio, Andrea Derdzinski, et al. Circumbinary Disks: Accretion and Torque as a Function of Mass Ratio and Disk Viscosity. *The Astrophysical Journal*, 901(1):25, September 2020. 8

[58] Dennis B Bowen, Vassilios Mewes, Scott C Noble, et al. Quasi-periodicity of Supermassive Binary Black Hole Accretion Approaching Merger. *The Astrophysical Journal*, 879(2):76, July 2019. 8

[59] Maria Charisi, Zoltán Haiman, David Schiminovich, and Daniel J D'Orazio. Testing the relativistic Doppler boost hypothesis for supermassive black hole binary candidates. *Monthly Notices of the Royal Astronomical Society*, 476(4):4617–4628, June 2018. 8

[60] Daniel J D'Orazio, Zoltán Haiman, and David Schiminovich. Relativistic boost as the cause of periodicity in a massive black-hole binary candidate. *Nature*, 525(7569):351–353, September 2015. 8

[61] G Hallinan, V Ravi, and Deep Synoptic Array team. The dsa-2000: A radio survey camera. *Bulletin of the AAS*, 53(1), 1 2021. https://baas.aas.org/pub/2021n1i316p05. 8

[62] Eric J Murphy, Alberto Bolatto, Shami Chatterjee, et al. Science with an ngvla: The ngvla science case and associated science requirements. *arXiv preprint arXiv:1810.07524*, 2018. 8

[63] Peter E Dewdney, Peter J Hall, Richard T Schilizzi, and T Joseph LW Lazio. The square kilometre array. *Proceedings of the IEEE*, 97(8):1482–1496, 2009. 8

[64] G Janssen, G Hobbs, M McLaughlin, et al. Gravitational Wave Astronomy with the SKA. In *Advancing Astrophysics with the Square Kilometre Array (AASKA14)*, page 37, April 2015. 8

[65] Rendong Nan, Di Li, Chengjin Jin, et al. The five-hundred-meter aperture spherical radio telescope (fast) project. *International Journal of Modern Physics D*, 20(06):989–1024, 2011. ⤴

[66] Lei Qian, Rui Yao, Jinghai Sun, et al. FAST: Its Scientific Achievements and Prospects. *The Innovation*, 1(3):100053, November 2020. ⤴

[67] Na Wang. Xinjiang qitai 110 m radio telescope. *Scientia Sinica Physica, Mechanica & Astronomica*, 44(8):783–794, 2014. ⤴

[68] MT Lam. Optimizing pulsar timing array observational cadences for sensitivity to low-frequency gravitational-wave sources. *The Astrophysical Journal*, 868(1):33, November 2018. ⤴

[69] KJ Lee, CG Bassa, GH Janssen, et al. The optimal schedule for pulsar timing array observations. *Monthly Notices of the Royal Astronomical Society*, 423(3):2642–2655, July 2012. ⤴

Index

Note: Locators in *italics* represent figures and **bold** indicate tables in the text.

Printed in the United States
by Baker & Taylor Publisher Services